Managing international technology transfer

"To survive in the future a firm . . . must be able to transfer technology to other countries better than the competition."

R.T. Keller and R.R. Chinta

As a result of the rapid globalization of business the subject of international technology transfer is now of vital importance to both scholars and practitioners of international business.

In drawing his examination from the fields of international studies, international management and international law Min Chen provides the first comprehensive consideration of the management and legal implications of international technology transfer. The book is divided into four organically linked parts which discuss comparative national policies on technology transfer, the international protection of intellectual property, international technology licensing and, lastly, international technology transfer via other major commercial channels.

This book will be of particular interest to students and academics in business and technology, also to international executives and lawyers.

Min Chen is Assistant Professor at the American Graduate School of International Management, Arizona, USA.

Thunderbird series in international management

The American Graduate School of International Management, Thunderbird, focuses on global business education. Combining strengths in all the key areas of international management, it provides a multicultural and multilingual approach to doing business worldwide.

The series presents a range of books which look at the various aspects of international management. Its breadth of coverage ranges from cultural analyses for effective management practice to key issues in world business. Designed to represent the variety of the work undertaken by the school at all levels, whether research into specific topics, textbooks in core business functions, or books directed to the thinking manager, the overall aim is to present the importance of international management issues to an international audience.

Managing international technology transfer

Min Chen

INTERNATIONAL THOMSON BUSINESS PRESS
I(T)P An International Thomson Publishing Company

London • Bonn • Boston • Johannesburg • Madrid • Melbourne • Mexico City • New York • Paris
Singapore • Tokyo • Toronto • Albany, NY • Belmont, CA • Cincinnati, OH • Detroit, MI

Managing International Technology Transfer

A division of International Thomson Publishing Inc.
The ITP logo is a trademark under licence

British Library Cataloguing in Publication Data
A catalogue record for this book is available from the British Library

Library of Congress Cataloguing in Publication Data
A catalogue record for this book is available from the Library of Congress

---

First printed 1996

Typeset in the UK by J&L Composition Ltd, Filey, North Yorkshire
Printed in the UK by TJ Press (Padstow) Ltd, Padstow, Cornwall

---

ISBN 0-41513-323-8

International Thomson Business Press
Berkshire House
168–173 High Holborn
London WC1V 7AA
UK

International Thomson Business Press
20 Park Plaza
14th Floor
Boston MA 02116
USA

Contents

Figures

Tables

Preface

In recent years, academic interest in international technology transfer has grown enormously. Studies in this area examine the economic relationship between a technology supplier and a recipient as well as a whole series of related issues, such as the relevant national policies and the legal framework for technology transfer in the various nations of the world. Although technology transfer across national boundaries had already begun to occur long before the Industrial Revolution, it did not emerge as a separate field of research until the 1970s, when people began to appreciate the ever-deepening interdependence of nations and the importance of technology in promoting the competitiveness of their national economies.

For highly diverse environmental, cultural and developmental reasons, technological advances in different countries have always been uneven. This tendency toward uneven development, central to technological progress in the world, serves as the basis for international technology transfer. In the post-World War II era, the uneven development of technologies has accelerated; consequently, technology transfer across national boundaries has experienced phenomenal growth. Indeed, the growth rate of technology transfer has been well ahead of that of the tangible goods trade and capital export. And yet, study of this phenomenon has lagged far behind the rapid developments taking place in the business world.

Until now, international technology transfer has been studied in a fragmented way by scholars from several separate fields. Each of these groups has focused on aspects of the phenomenon that are only of interest to that group. Thus far, most of the literature has been written by professors of international studies or political science (focusing on policy issues), professors of international management (focusing on transnational management) and lawyers and law professors (focusing on legal intricacies of international transfer).

This book represents an attempt to combine these different approaches into an organically linked volume on international technology transfer. It is designed to provide readers with a comprehensive approach and with a systematic education on key aspects of the subject. It examines such issues as national policies on technology transfer and foreign direct investment, the complexities of protecting intellectual property, the challenges of licensing technologies to foreign recipients and negotiating technology prices, the controversial issue of restrictive business practices, the fundamentals of transferring technology via other major commercial channels, and an introduction to dispute resolution mechanisms. The subjects covered in the book will provide the reader with a solid foundation on how to manage international technology transfer successfully.

The book comprises an introduction, four parts of main text, and a conclusion. The introductory chapter lays out a conceptual framework, covering definitions, theoretical discussion, and a review of the relevant literature; this chapter also provides a preview of the book's organizational structure.

Part I compares national policies on technology transfer in terms of both acquisition and export of technology, as well as technology export control issues. Since technology can be transferred through various forms of foreign direct investment, Part I also compares national foreign direct investment (FDI) policies of various host governments. Part II is an introduction to intellectual property protection and its relevancy to international technology transfer. This part of the book includes a conceptual introduction to intellectual property rights (IPR), a survey of international organizations involved in IPR protection and an analysis of existing gaps in IPR protection between developed and developing countries, and concludes with a discussion of the challenge of enforcement. Part III concentrates on international technology licensing, price negotiation, contract formation and the issue of restrictive business practices. Part IV explores technology transfer via other major commercial channels, with an emphasis on joint ventures, international subcontracting and dispute resolution mechanisms.

Finally, the concluding chapter of the book examines the dynamic relationship between international business and international technology transfer, assesses the costs and benefits of international technology transfer, analyzes major existing problems in this field, and makes a few tentative suggestions for future improvement in the practice and study of international technology transfer.

This book is mainly intended for international executives and business school students who have an interest in doing international business. Therefore, the message is conveyed in a straightforward style with relatively plain language. Still, the book may be of interest to those academics who are involved in teaching and research in the area of international technology transfer. For that reason, the book provides brief reference notes throughout and lists suggestions for further reading at the end of the book. (All monetary units are denominated in United States dollars, unless otherwise specified in the text.)

Writing a book such as this proved to be a time-consuming, complicated process. I could not have accomplished this task without support from many people. In this sense, it was a collective effort as much as an individual endeavor. I am deeply indebted to my colleagues in the Department of International Studies at Thunderbird, the American Graduate School of International Management (AGSIM), particularly the Department Chair, Dr Llewellyn D. Howell, and Professor Robert Tancer. Without their constant support and encouragement, this book would not have been completed in such a timely manner.

I wish to acknowledge Director Kang Rong-ping, of the Beijing Iron and Steel Corporation's Multinational Business Research Institute, for his contribution to Chapter 1. Many thanks go to Professor Richard D. Robinson who kindly permitted the use of materials from his case book. Mr Geoff Martin, a friend and graduate assistant, spent considerable time in helping to edit the book and deserves profound thanks for his dedication. I owe a special debt to my research assistant, Ms Linda H. Liao, who provided many good suggestions on various parts of the book. Thanks are also due to the graduate students in my classes on international technology transfer and foreign direct investment, whose comments gave me very useful insights into the subject matter.

I also wish to thank my parents, whose international careers and examples as role models have had tremendous influence on my later academic pursuits. My family deserves my special thanks for their consistent encouragement and support. Without their understanding, this book would never have been finished.

I hope that this book will provide international business executives from both developed and developing countries with a better understanding of the dynamics of international technology transfer. Such an understanding provides a crucial foundation for cooperation between technology transferors and transferees. In this way, it can contribute greatly to the success of international technology transfer transactions.

Min Chen
Glendale, Arizona

Introduction

Concepts and issues

INTRODUCTION

International technology transfer studies cover the economic relationship between a transferor and a transferee as well as a whole series of related issues, such as the relevant national policies and legal framework of the nations in the world. Owing to different environmental and developmental reasons, technological advances in different countries have always been uneven. This uneven nature of technological progress throughout the world provides the very basis for technology transfer. In the past few decades, international technology transfer has multiplied by leaps and bounds.

According to recent statistics of the United Nations, the total technology trade of the world amounted to $3 billion in 1965, reached $11 billion in 1975 and surpassed $40 billion in 1985, a growth rate of thirteen times in twenty years. In the past twenty years, this growth rate has been well ahead of that for ordinary merchandise trade and capital export (Li 1993: 7). Technology trade has not only formed an independent market, but also become an extremely important part of international economic relations. Export of technology and relevant experience has become a distinct trademark of a developed country.

DEFINING TECHNOLOGY

"Technology" is a difficult concept to define. The concept of technology has been defined in many ways and from different angles (Stewart and Nihei 1987). Simply put, it refers to a class of knowledge for making a specific product. The technical skills necessary to utilize a production technique and a product are often included in the definition of technology. The *Licensing Guide for Developing*

Countries prepared by the World Intellectual Property Organization (WIPO) has provided a fairly comprehensive definition:

> Technology means systematic knowledge for the manufacture of a product, the application of a process or the rendering of a service, whether that knowledge be reflected in an invention, an industrial design, a utility model, or a new plant variety, or in technical information or skills, or in the services and assistance provided by experts for the design, installation, operation or maintenance of an industrial plant or for the management of an industrial or commercial enterprise of its activities.
>
> (WIPO 1977: 28)

It is very important to emphasize the role of technology in the economic arena. As Stewart and Nihei (1987: 1) state, "Technology refers to new and better ways of achieving economic ends that contribute to economic development and growth."

As a commodity, technology has a number of unique features. First, it does not have a fixed shape, consisting mainly of design, documents and prescriptions. It can also be diffused through oral dictation and illustration. Second, it does not need reproduction to be used and transferred multiple times. Therefore, its marginal cost is almost zero. Third, technology transfer often refers to the transfer of the right to use instead of the right to ownership. Therefore, it has a public good nature. When knowledge is given to another party, that knowledge usually remains available to the transferor. The benefits of using a technology can be seriously influenced by the number of people having the right to use it.

Every specific technology has a life cycle characterized by the S-curve, which describes the maturation and replacement of a technology as well as the necessity to implement different strategies for a technology at different stages of maturity (Figure I.1). In the embryonic stage (development and application recognition), the technology, which has just passed experimental requirements, is not used for large-scale production and the value of transfer is not normally high. When a technology grows into the mature stage (application launch and application growth), the value of transfer is climbing to its climax, as it generates increasing economic benefits to the producers. But as a technology moves into its aging stage (technology maturity and technology degradation), it reaches its ceiling of performance. The marginal cost of developing an additional improvement grows while the value of transferring this technology

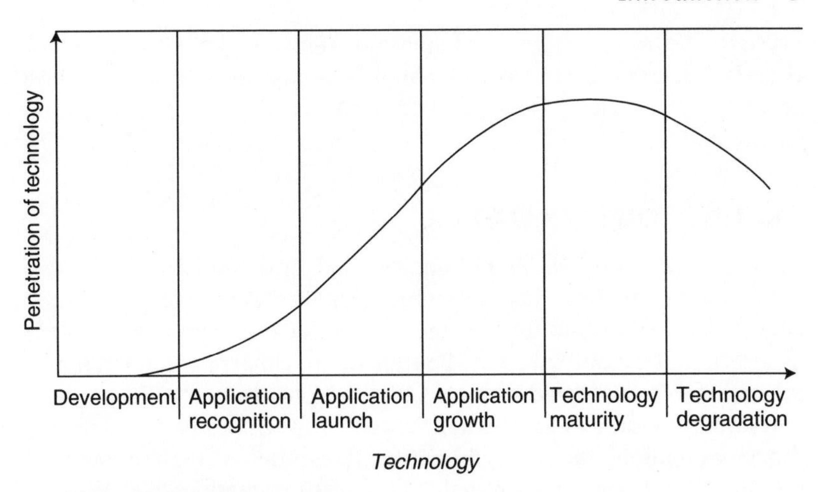

Figure I.1 Technology life cycle
Source: Frankel 1990: 72 (reprinted by permission of Kluwer Academic Publishers)

declines. The technology begins to experience a downturn and degradation, and eventually it is abandoned.

Technology has a diversity of forms, ranging from a fairly simple agricultural process to very complex computer systems. A distinction may be made between "conventional technology" and "high technology." Conventional technology comprises, for instance, the technology in the textile, paper and cement industries. In contrast, high technology industries are characterized by heavy research and development expenditure, rapid and continual technological change, the important role of complex patented or non-patented proprietary technology and large capital requirements (Chudson 1971: 19).

There is also the distinction between open, semi-open and secret technologies. Open technology refers to various published and circulated technological theories and practical knowledge, such as scientific and technological theses, academic reports and open technology forums. Semi-open technologies refer to patented technologies. Inventors register their inventions for patent protection, under which nobody can use the invention without the permission of the patentee within the effective period of a patent, otherwise that person infringes on the right of the patentee. On the other hand, the patentee should publicize his technology so that the public can understand it and follow legitimate ways of using this technology. Inventors usually do not publicize all of the contents of their technology but keep at least some core parts secret. Therefore, patented technology becomes semi-open. Secret technology refers to know-how, which consists of unpublicized core technologies and data, which are best protected by security measures and the laws on protecting commercial secrecy.

TECHNOLOGY TRANSFER

The term "transfer" is more controversial in definition. People disagree on the factors that determine whether a transfer has really occurred. Some argue that technology is not transferred unless it is absorbed and actually used by the transferee. Others contend that how the transferee deals with the transferred technology should not be a determining factor as to whether the technology is in fact transferred. Since technology is intangible knowledge rather than a tangible product, the definition of transfer is hard to operationalize. When a product is transferred across a national boundary, it no longer has a physical presence in its original place. Even if it is not used, the transfer has already taken place. By transferring knowledge, the transferor has not automatically given up the knowledge, but rather shares it with the transferee. Therefore, it is not unreasonable to say that a transfer is not achieved until the transferee understands and can utilize the technology (Erdilek and Rapoport 1985: 252).

The definition of transfer can be further complicated by the diverse channels through which it can occur. There is no one best way of transferring technology for two reasons: first, technology does not exist in a social vacuum, but rather is "embodied in products, processes and people"; second, technology circulates through very diverse institutional channels or mechanisms (Goulet 1978: 32). There are both formal and informal channels, of which some involve voluntary and intentional international technology transfer and others do not.

The principal channels of international technology transfer are licensing; franchising; direct foreign investment; sale of turnkey plants; joint ventures, subcontracting, cooperative research arrangements and co-production agreements; export of high-technology products and capital goods; reverse engineering; exchange of scientific and technical personnel; science and technology conferences, trade shows and exhibits; education and training of foreigners; commercial visits; open literature (journals, magazines, books and articles); industrial espionage; end-user or third country diversions; government assistance programs, etc.

Technology transfer through most of these channels is very difficult to monitor. Through formal channels (the first six above), technology is transferred via a market mechanism and has an explicit value. Even when an explicit value is available, it may not reflect the "true" value of the technology being transferred. Moreover, if actual transfers cannot be clearly recognized through informal channels, it is almost impossible to determine the more important transfer channels (Erdilek and Rapoport 1985: 252–253). To make research manageable, this book applies a more narrow definition (as in formal channels) which defines technology transfer as "a process by which expertise or knowledge related to some aspect of technology is passed from one user to another for the purpose of economic gain" (Schnepp *et al.* 1990: 3).

There are a number of parties interested in international technology transfer. National governments are concerned with how international technology transfer affects domestic economic development, international competitiveness, and national security. They are interested in the need for and the usefulness of national control on either outflows or inflows of technology. Between 1955 and 1970, with the help of its industrial policy, Japan absorbed almost all the advanced technologies invented in the first half of the twentieth century and successfully modernized its economy, becoming the second largest economic power. Japan spent only US$6 billion; if the Japanese had had to invent this technology, they would have had to spend US$180 to 200 billion (Li 1993: 7).

Several international institutions, including the General Agreement of Tariffs and Trade (GATT), United Nations Conference on Trade and Development (UNCTAD), United Nations Industrial Development Organization (UNIDO), the United Nations itself, and the Organization for Economic Cooperation and Development (OECD), have set up commissions and study groups to investigate various aspects of the subject. In the past two decades, several UN organizations have been particularly interested in the impact of North–South technology transfer.

Corporations and individuals who directly participate in international technology transfer as either suppliers or recipients of technology are undoubtedly also interested in a better understanding of the field. Multinational corporations are major vehicles of international transfer of technology (Rugman 1983). Their activities cover the selection of technologies for transfer, the choice of transfer channels, and transaction pricing and payment mechanism negotiation. In addition, they have to deal with the attempts of national governments to impose various restrictions on international technology transfer.

There are a number of unique features in technology transfer. First, commercial technology transfer is highly monopolistic, since technology as the product of an invention is unique. The contemporary patent system has further reinforced such monopoly. In order to maintain the advantage of its technology and products, the owner of a technology does not normally transfer the technology, except in some specific situations – for example, when a transfer is necessary for occupying the market, when the transfer can bring huge profits, or when the transfer does not threaten its monopoly. In addition, as a result of unequal development of science and technology, developed countries tend to impose various restrictions on the transfer of advanced technology in order to maintain their monopolistic position.

Secondly, technology transfer can be of multiple exchange. One technology can be traded multiple times, as the transfer does not involve ownership but only the right to use. The number of transfers will have a direct impact on the value of the technology. When a technology is transferred to a variety of end-users, the price the transferor charges will drop correspondingly. Only in specific situations is ownership transferred, but the price is usually very high and conditions are very rigid. In contrast, ordinary merchandise trade involves the transfer of both ownership and the right to use.

Finally, technology transfer does not simply follow the basic market rule of exchange. The price of the technology is not simply determined by its value (i.e., research and development (R&D) and profits), but by the profits it can bring, i.e., the total profits that the transferee can achieve by utilizing this technology. The price of the technology transferred is the "licensor's share of licensee's profits." When the transferor is in a monopolistic position, he will tend to charge high prices. When he is not in a monopolistic position, he will have to settle on a price for the technology that is far lower than the price needed to recapture the full cost of the technological effort, simply because the marginal costs associated with an additional sale are usually very low and the costs of generating the technology itself are sunk (Vernon 1986: 48).

INTERNATIONAL PRODUCT LIFE CYCLE THEORY AND TECHNOLOGY TRANSFER

A closely related concept to the technology life cycle is the product life cycle. Products have life cycles encompassing initial innovations, incremental innovations, and a declining process to the end. In a product life cycle, innovation is initially concentrated on product change, and later shifts its focus to process change. If an industry is based on a single product line, then the technology life cycle and product life cycle coincide. But if an industry comprises several product lines, the technology life cycle can be compared to "the envelope of the several product life cycles" (Betz 1987: 74).

Aggarwal (1991: 64–68) has provided a detailed analysis of the relationship between product life cycle and technology transfer across national borders by presenting a version of the international product life cycle based on three categories of countries: the technologically advanced home country where the product is originally developed; a group of countries that may be at the same or somewhat lower technological level and may be classified for convenience as the more developed countries (MDCs); and a group of countries that are at a lower level of technology and economic development and are commonly called the less developed countries (LDCs).

According to the international product life cycle theory of technology transfer, the international participation of a business can be examined by following sequential stages in the life cycle of the product or process it develops. Owing to the restrictions in the flow of information across national borders and the growing demand for high-income consumer goods and labor-saving producer goods, innovation of new products is far more likely to be realized in highly developed and industrialized nations. These same innovations are more likely to be applied in less developed nations later as these nations develop economies and consumer tastes similar to those of the highly industrialized nations.

Without denying the fact that new product innovation can and does occur in all countries at all levels of economic development, Aggarwal for the sake of simplicity concentrates his analysis only on the cases where new product innovation is developed in the United States. He argues that the analysis is also applicable to cases where new product innovation takes place in countries other than the United States.

According to his analysis, the first phase of product development and domestic sales growth seems to be critically important in the life of the product. It is during this period that the product undergoes test marketing, product redesign and re-engineering, and production scale-up. Only when a product wins wide acceptance in the domestic market is there any necessity to look into international technology transfer. Under normal circumstances, if a product does not get domestic market acceptance, it will very likely not be exported. But if the new product is accepted domestically and at this first stage domestic sales volume enjoys growth, the product may move to the second phase.

During the second stage, domestic sales continue to grow, but the growth rate may begin to decrease. Exports to other MDCs may take place; and some of them may even go to LDCs. Up to this stage, however, all production still remains in the United States. The product may be exported to MDCs for three basic reasons:

1 Similarities between the markets in the MDCs and the United States give rise to demand for the product.
2 The price for the product has been reduced due to economies of scale that have developed along with market growth.
3 The economic level of the MDC may have increased.

During this stage, technological changes also occur in the MDCs that actually usher in the next phase. As MDC residents' expertise on this product is promoted, their ability to establish a manufacturing operation in their country is very likely to improve correspondingly. At this juncture, the industry experiences heated price competition at home because other domestic firms have developed similar or substitute products that are being sold in the American market and in the markets of the MDCs. Some MDC governments may eventually raise tariffs or impose quotas to entice the US firm to invest in their countries. Consequently, the US firm may be obliged to invest in one or some of the MDCs – the third phase of the life

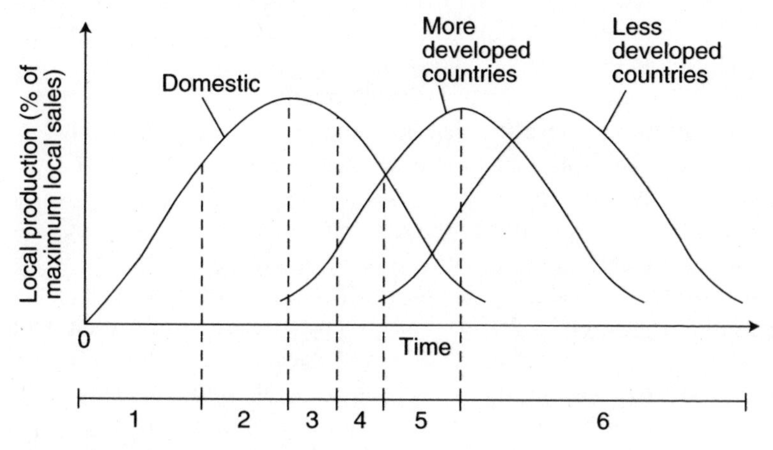

Figure I.2 Product life cycle phases
Source: Aggarwal 1991: 67

cycle of the product. The US firm may also have to make the decision to transfer technology.

In the third phase of the cycle, when production has started in a foreign market, American exports to that market may no longer grow as rapidly as before, or may even begin to experience decline. At the same time, American exports to non-producing countries may begin to be displaced by exports from other producing countries. When overseas markets grow to a substantial size, MDC manufacturers will no longer suffer from the high costs associated with small-scale production. In many cases, the MDC-producing company may be a subsidiary of an American parent.

In the fourth phase, these firms may take away some markets in LDCs that were previously supplied directly from the United States. In the fifth phase, foreign production in some countries reaches such a large scale that costs become low enough to offset the transportation and tariff protection of the US manufacturer and the United States becomes a net importer of the product. In the last phase, the LDCs' manufacturers, with lower labor costs and improved skills, become exporters to the MDCs and the United States.

For Aggarwal, one important implication of the product life cycle theory of technology transfer is that investment in R&D can be recovered over a much longer product life. This may also mean that an LDC may find it cheaper to buy new technology rather than develop it locally, because an MNC can spread the R&D costs across a worldwide market. Moreover, by the time a technology is transferred to LDCs, the company is likely to have recovered the major portion of the R&D costs and thereby may be willing to price the technology only on the basis of its much smaller marginal cost. Therefore, both sides may benefit from technology transfer.

LITERATURE REVIEW

International technology transfer did not emerge as a separate field of research until the 1970s as a result of the accelerating awareness of the economic interdependence of nations and the key role of technology in economic development. Even in the 1970s, there were only a limited number of centers focusing on the research of technology transfer (Jolly and Creighton 1975). The study of international technology transfer is prompted by both the policy considerations in the public sector and the need in the private sector to better understand the process itself.

The aspects of international technology transfer that have been research priorities generally fall into three broad categories: understanding the process itself, particularly its determinants; effects on transferor and transferee; and factors affecting its control. Although one can debate whether these issues are any more significant now than they used to be in the past, they are currently attracting substantial interest. If the concept is more clearly defined and international technology transfer can be measured in more sophisticated ways, some of these important issues can be better examined and resolved (Erdilek and Rapoport 1985: 251).

Since the mid-1970s, the field of international technology transfer has produced a rapidly growing and heterogeneous body of academic literature. Policy-oriented economists, management specialists, and lawyers have made major contributions to this burgeoning literature, followed later by political scientists, international relations specialists, and sociologists of science, technology and development. While they have different focuses, two main streams of research can be identified (Contractor and Sagafi-nejad 1981: 113–135; McIntyre 1986: 13–17). One concentrates on the technology recipient side, the other on the technology supplier side of the transfer equation.

Based on the assumptions of the "dependencia" school of economic development, the first stream believes that technology transfer is conducted in a seller's market and that the free market does not provide a very appropriate mechanism. International codes and systems through which technology suppliers and receivers can find a common ground to reconcile divergent interests are necessary. With an emphasis on states as units of analysis rather than companies, this stream of research has focused on a number of research topics:

1 The effects of technological oligopoly of a few firms and countries on recipients' economies.
2 The role of patents, trademarks, copyrights, and business practices in maintaining such oligopoly and in the protection of intellectual property.
3 The appropriate modes of transfer and usefulness of the transferred technology.
4 The fair pricing of technology transfer.
5 The policies of recipient governments to either facilitate or restrict the transfer.
6 The effects of supplier countries' policies on the transfer process.
7 The impacts of corporate strategies on the recipient countries.
8 The endeavors at creating multilateral policies in the form of codes.

In contrast, the second stream tends to be more concentrated on the technology supplier side of the transfer equation. Research has been largely done from the perspective of the supplier firm (e.g., a multinational corporation (MNC)), especially in the areas of strategic management, corporate policy, and organizational behavior. The focus has been narrower, with more emphasis on corporate policy than on public policy. The dominant assumptions of this stream are that the technology market is a buyer's market and the market mechanism provides a much better solution than any international or national regulating system. This stream, strongly influenced by liberal economic ideals, typically advocates the view that integration of national economies, to the extent of the internationalization of production, is the best way to promote global economic efficiency. Three core issues have been dealt with:

1 The managers' decision-making process in selecting the technology package to export.
2 The decision-making process of multinational corporations in selecting the proper channels of technology transfer to maintain control over R&D and manufacturing while return is maximized.
3 The suitable ways in which multinational corporations manage their relationships with overseas subsidiaries, other companies and governments that depend on their technology.

Although much work has been done in the existing literature, there still exist a number of problems. First and foremost, there is no organic multi-disciplinary approach in studying international technology transfer, which is itself a multi-disciplinary subject. Most scholars have focused on one or two aspects of technology transfer, be it legalistic, economic, political, or managerial. Therefore there is a lack of an integrative theoretical framework for the study of international technology transfer (Contractor and Sagafi-nejad 1981: 128).

Second, the existing literature has been so far overwhelmingly focused on technology transfer from North to South while having paid little attention to South–South technology transfer and North–North technology transfer. In fact, North–North technology transfer has been a very important part of trade and investment relations among the developed countries. And South–South technology transfer has gained substantial ground in the past decade. Such countries as China and India, main recipients of advanced technology themselves, have become major sources of technology for many less developed countries.

Third, the fact that the two schools take sides has left the literature unnecessarily fragmented. As will be shown, technology transfer generates incremental profits that both sides will share. As in many other businesses, it takes two to tango. As technology is traded under an imperfect market situation, characterized by bilateral monopoly, any successful technology transfer must take the interests of both sides into consideration. Therefore, a relatively unbiased approach is needed for the study of technology transfer.

GENERAL LAYOUT OF THE BOOK

This book, written in an attempt to avoid those problems of bias, is divided into four parts. The first part focuses on the relationship between technology transfer and national policies. It includes discussions on major national acquisition models of technology in the world, an in-depth analysis on the Japanese technology export model, a comparative study of three categories of national policies toward inward foreign direct investment (FDI) and finally a general discussion on the evolution of US national control of technology export and its implication for technology transfer. This part is the primary concern of public policy makers, but provides an indispensable policy introduction to business practitioners.

The second part is an introduction to intellectual property protection and its relevancy to international technology transfer. The prominent features of such concepts as patent, trademark, copyright and know-how are discussed. The importance to international technology transfer of protecting these intellectual properties is highlighted. Various international conventions on registering and protecting these properties are introduced and compared. Major existing problems and prospects of overcoming them are explored. In the end, strategies to protect intellectual properties in the process of technology transfer are discussed. Although some discussions are highly legalistic, both policy makers and business practitioners also benefit.

The third part concentrates on international technology licensing. It first provides a theoretical discussion on licensing, comparing it with export and FDI in an attempt to bring out their comparative advantages and disadvantages under different circumstances so that business practitioners can make proper decisions. Following this discussion is a chapter on technology pricing negotiation. The positions of both licensors and licensees are explored. This is followed by a detailed analysis of a licensing agreement. Finally, the complicated issue of restrictive business practice in international technology transfer and various international and national policies of controlling such practices are discussed. This is the core area of managing technology transfer.

The fourth part explores technology transfer via other commercial channels. It begins with a comparison of other commercial arrangements for technology transfer, including franchising, subcontracting, turnkey projects, counter-trade; R&D cooperations and technology swaps, joint production agreements, equity joint ventures, joint equity swaps; and intrafirm transfer. Joint venture as a major channel for technology transfer is analyzed in detail in the context of a developing country (China). This is followed by a discussion on technology transfer and international subcontracting. Finally, various dispute resolution mechanisms for foreign direct investment are introduced.

With regard to the organization of discussions on technology transfer, the author tends to agree with the basic model created by Samli (1985), which keeps five key components in balance: the sender, the technology, the receiver, the aftermath, and the assessment. It also includes six dimensions of technology: geography, culture, economy, people, business and government. In other words, the author favors neither the supplier nor the recipient. For this author, the process of good technology transfer should be characterized by mutual benefit and involve equal efforts on both sides for success. Technology transfer is extremely complex, encompassing many dimensions. Some of these dimensions may work against

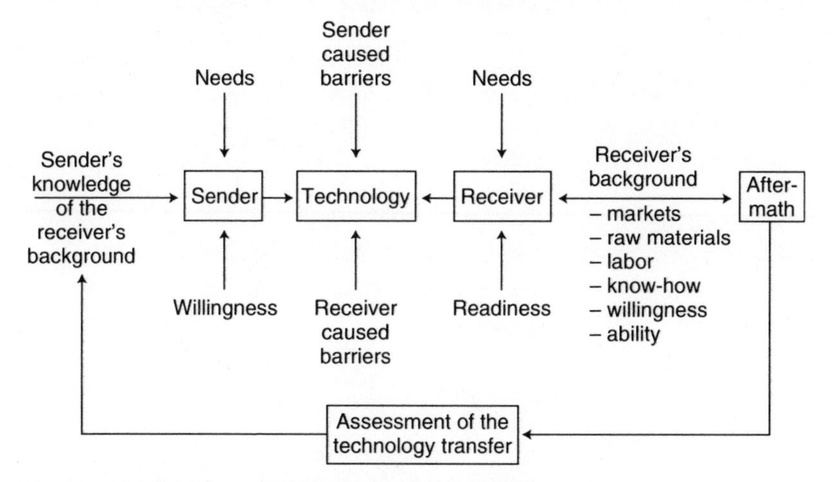

Figure I.3 The basic model of technology transfer
Source: Samli 1985: 9 (reprinted by permission of Greenwood Publishing Group, Inc., Westport, CT)

each other. Therefore, any discussion on the process of technology transfer should be based on a comprehensive approach, which includes not only the technical aspects of the transfer but also the related environmental factors.

SUMMARY

This introductory chapter has laid down a conceptual framework for the whole book. It first defines technology, with an emphasis on the role of technology in the economy, then proceeds with the definition of technology transfer. Once again, technology transfer in this book does not cover all sorts of transfer that can take place in the world, but rather focuses on the kinds of transfer that would generate economic gains for the parties involved. Following this is a brief discussion of the international product life cycle theory and its relationship to international technology transfer. Then a brief critique of existing literature is conducted and main problems are identified. In spite of the thriving literature in a young field like technology transfer, there is still a lack of an integrative theoretical framework and studies have been fragmented. Finally, at the end of the chapter, the layout of the book is explained.

Part I

Comparative national policies

1 Comparative national models of technology acquisition

INTRODUCTION

Technology acquisition strategies vary a great deal in different countries and national policies on technology acquisition have constantly changed in the course of economic development. Different cultural, historical and developmental environments have impacted various national strategies and produced different policies on foreign technology acquisition. A hasty attempt at generalization could easily distort the true picture. In the course of development, different countries do have identifiable thrusts, with each thrust having specific advantages and disadvantages.

In this chapter, these varied thrusts are generalized into several generic models, with which one can make a preliminary assessment of the advantages and disadvantages of various national policies on technology acquisition. These basic models include: the migrant model, which focuses on attracting talents from other countries; the hardware model, which places emphasis on the large-scale acquisition of industrial equipment; the software model, which is based on the acquisition of technology information (licensing); and the capital model, which attaches great significance to the attraction of foreign direct investment (FDI). In general, laboratory technology licensing represents the highest level of technology acquisition while the migrant model has the best overall effects on technological and economic development. On the other hand, the hardware model probably is the most expensive and least effective in the long run. This chapter focuses on discussions and comparisons of these technology acquisition models.

MIGRANT MODEL

The model of technology acquisition promoting the immigration of skilled talents from other countries is called the migrant model. The United States of the eighteenth and nineteenth centuries is a most typical example. This country proclaimed political independence in 1776, accomplished technological independence a century later and eventually became the foremost technological power in the world, providing the largest source of the most advanced technologies.

The United States is by far the leading nation of immigration in the world. Between 1820 and 1986, the recorded numbers of immigrants exceeded 53 million. Until about 1880, most immigrants to the United States came from Northern and Western Europe, especially England, which was technologically the most advanced nation at the time. Later on, immigrants from Southern and Eastern Europe poured into the country. During the Second World War, German fascists persecuted many ethnic groups in Europe, and a large number of the scientific and technological elite subsequently escaped to the United States. Many of those immigrants later became the backbone of science and technology in the United States. Among the most well known of this group was Albert Einstein who formulated the theory of relativity. In the postwar immigration tides, most new immigrants have come from developing countries. A lot of these new immigrants were students who stayed in the United States after receiving their higher degree there. Currently, the United States annually absorbs well over half a million documented immigrants from all over the world (excluding illegal immigrants). As President John F. Kennedy once wrote, the United States is "a nation of immigrants" (Gardner and Bouvier 1990: 341–362; and Jones 1992).

The transfer of European technology to North America constituted the basic process of the American industrial revolution. From the beginning of the North American colonies until the nineteenth century, North America mainly depended upon England and other West European nations for technology. European immigrants played a very important role in the process of technological progress in the United States. Each of the formative industries in the United States – textiles, mining and iron and steel – relied heavily on immigration of European, especially British, artisans, operatives and entrepreneurs. As the American economic historian Nathan Rosenberg commented:

> The rapid rate of technological change reflected both a high level of inventive activity and a rapid rate of adoption, the advantage of all

latecomers of being able to borrow and to modify a technology that had been developed by others. Immigrants to the United States brought European technology with them and continued to draw heavily upon it.

(Rosenberg 1972: 32)

The role played by skilled immigrants in the process of technology transfer was crucial to the development of the United States.

The adoption of the migrant model by the United States can be explained in several ways:

1 As the largest immigrant country, the United States was naturally suited for this model.
2 Modern technology originated in Western Europe, with Great Britain in the lead. Having close cultural affinity with European nations and even sharing the same language with Great Britain, the United States could easily absorb these immigrants.
3 For an extended period, Great Britain implemented a policy of prohibiting the export of new machinery and the great physical distance also made it difficult and costly to ship large machinery and equipment. Therefore, the United States had to depend on immigrants for new technology.

After the end of the Second World War, the United States gradually formed the strategy of attracting talents from other countries on the basis of its historical experience. Consistent preference has been given to those having good education and useful skills that the country needs. For example, foreign professionals are third on a list of seven preferences for desired immigrants, after sons and daughters of US citizens and spouses and unmarried sons and daughters of aliens lawfully admitted for permanent residence. This third category "is for members of the professions or persons of exceptional ability in the arts and sciences" (Danilov 1989: 11). Based on the ever-changing needs of the country, specific items on the preference list have varied a great deal in US immigration law for foreign professionals at different periods.

This immigration policy has saved the United States hundreds of thousands of dollars in education and training. In 1972 alone, the United States made savings of about $883 million from 16,012 skilled immigrants (Mandi 1981: 42). In the early 1980s, one-fourth to one-third of the design engineers in the United States were foreigners. With the rising competition in technology of the 1980s, the American ability to exploit talents from other countries was one of its greatest competitive advantages (Nussbaum 1983: 222). No other nation is as open to the scientists and engineers of other countries and no other nation has benefited as much. Two major barriers hindering many countries from taking advantage of migrant talents are cultural and ethnic; the openness of the American culture and the tradition of the melting pot has helped overcome these barriers.

Countries in the migrant model category include Canada, Australia, New Zealand, South Africa and some South American nations. Most of those countries have enacted policies and regulations to encourage immigrants who have useful skills and good educational backgrounds. In the postwar international migration tides, Australia, Canada and New Zealand have been among the most favored destinations for professional people from developing countries. While better living standards and working conditions in those countries have been the major reasons for immigration, various incentive policies have greatly accelerated this global migration movement.

In the case of Canadian immigration laws, for example, each applicant seeking admission to Canada as an independent applicant or an assisted relative is graded on a points system. The categories for admission are a complex attempt to acquire immigrants who are well trained in specific occupations, and the longer the training the more points an applicant will receive. There are ten categories in the points system:

1 Education.
2 Specific vocational preparation.
3 Experience.
4 Occupational demand.
5 Arranged employment and designated occupation.
6 Proposed location of living.
7 Age.
8 Knowledge of English and French.
9 Personal suitability.
10 Relatives in Canada.

(Segal 1981: 28–34)

HARDWARE MODEL

The model of technology transfer which places emphasis on the purchase of industrial equipment is called the hardware model. Since the beginning of the twentieth century, this model has become widely adopted by the late developing countries. The most typical and

Table 1.1 The percentage of industrial equipment in the total import of the former Soviet Union

Year	*1925*	*1929–30*	*1931*	*1933–37*	*1940*	*1950*	*1960*	*1965*	*1976–79*	*1980*
%	19.8	37.0	54.0	32.0	32.4	21.5	29.8	33.4	38.6	33.9

Source: Adapted from Soviet Foreign Trade Annual Statistics, 1925–80

pioneering form of this model was the former Soviet Union from the 1920s to the 1980s. In the course of the economic development of the developing countries, technology acquisition has played an exceptionally important role. In the post-Second World War period the hardware model once exerted significant influence on many newly independent countries.

For the seven decades of its existence, the former Soviet Union deviated very little from its main method of technology acquisition. From 1925 to the early 1980s, the importation of industrial equipment was maintained at a relatively high level (Nove 1984). Throughout most of its history, the former Soviet Union was one of the largest purchasers of machinery and equipment in the world. More than 50 per cent of industrial equipment exports in 1931, for example, were purchased by the former Soviet Union.

The purchase of complete sets of equipment was dominant in the Soviet acquisition of foreign industrial equipment. This pattern was obvious in the climax of technology acquisition in the 1930s, and remained so in the 1960s and 1970s. After the end of the Second World War, Soviet technology acquisition developed more in the direction of "pure" hardware purchases. Acquisition of talents, which was fairly substantial in the 1930s, was limited to a minimum after the War and foreign investment was virtually forbidden. Technology licensing, which did not occur until the 1960s, remained at a very low level. Meanwhile, complete sets of equipment continued to occupy a large share of Soviet purchases. From the comparisons in Table 1.2, we can see the unique pattern of Soviet technology transfer, which had a higher percentage of equipment import than not only the world average but also the average figures for developing countries. This was in sharp contrast to the typical software model of Japan.

During the 1930s, the former Soviet Union was successful in quickly laying the foundation for industrialization via the large-scale acquisition of industrial equipment and technology. The speed

Table 1.2 The percentage of industrial equipment in the total import of different categories of countries

	1958	*1963*	*1970*	*1975*	*1978*
Former Soviet Union	29.0	33.4	35.1	33.9	42.0
Japan	11.2	10.2	11.3	6.2	6.9
Developed countries	16.8	20.2	26.9	24.6	26.1
Developing countries	28.6	29.0	33.5	35.2	35.8
World average	21.2	23.6	28.7	28.0	29.3

Source: Adapted from UN and Soviet Foreign Trade Annual Statistics, 1958–78

and scale of industrialization was unprecedented, though the quality was highly questionable (Jasny 1961). However, such a model had a very limited positive effect on Soviet development by the 1960s and 1970s. At the initial stage of industrialization, such a model may have an advantage of significant acceleration of technological advancement, especially applicable to the early stage of heavy industries which usually do not experience rapid technological change within a short period of time. But with the establishment of a national industrial foundation, the advantages of such a model decrease while its disadvantages increase. With the continued application of such a model, one can only maintain equal-distance pursuance at very high costs.

There are several major reasons for the former Soviet Union to adopt such a model. First, with the rapid development of big machinery production systems in modern times, movement of machinery and equipment as a way to realize technology transfer was both possible and necessary. The rich natural resources of the former Soviet Union enabled large-scale purchase of industrial equipment. Second, strategic and political factors were also very important, as the former Soviet Union was contained by major capitalist powers. After the Second World War, the coexistence of two world blocs (socialist and capitalist) and the influence of the Cold War further closed the former Soviet Union to Western countries, resulting in the persistence of hardware acquisition. Only through rapid industrialization could the Soviet Union survive the animosity of the major capitalist countries, and large-scale purchase of industrial equipment was almost the only option. Third, entirely different economic and political systems made FDI very difficult, if not totally impossible. Finally, the industrial heartland of the former Soviet Union was adjacent to European countries. This geographic approximation made the purchase of large industrial equipment relatively less costly.

From 1950 to 1985 China's technology imports basically followed the Soviet model. Due to a Western embargo on China throughout the 1950s, China imported technology and industrial equipment primarily from the former Soviet Union and Eastern European countries. During this period, 156 projects comprising a total of 400 items were imported at a cost of US$2.7 billion. The majority of these imports were completed sets of equipment from the Soviet Union that comprised 90 per cent of all foreign exchange spent on technology imports (Li 1985: 78–85). Generally speaking, this period was successful, as the large-scale introduction of technology and industrial equipment laid the basis for Chinese industries.

Beginning in 1960, the Soviet Union suspended all technology exports to China – a unilateral breach of contract. Consequently, China had to turn to Western Europe and Japan for technology, on a smaller scale until 1966. When the Cultural Revolution began in 1966, all technology import projects were halted and were not restored until 1973. From 1973 to 1978, China signed a total of 220 contracts worth US$3.8 billion, with twenty-six turnkey projects accounting for over US$3 billion (Conroy 1986: 22–24).

This was followed by a confusing period from 1978 to 1981, when radical economic development programs resulted in the sudden expansion of technology imports, regardless of efficiency or cost. During the sixth Five Year Plan (1981–85) over 10,000 medium to large projects were imported at a cost of more than US$10 billion. The importation of completed sets of industrial equipment still predominated although they were proportionally decreasing. After 1985, China began to emphasize licensing and FDI as a major channel to acquire foreign technology.

SOFTWARE MODEL

Technology transfer giving priority to the transference of technological information (licensing etc.) is called the software model. This is the model pioneered by postwar Japan (see Table 1.3). By the software model we mean that software acquisition costs constitute a relatively high percentage of the total costs of technology acquisition (including hardware) – 30 per cent or more in large-scale technology acquisition. As is shown in Table 1.3, Japan had close to 200 items of software acquisition in the early 1950s, the same level reached by South Korea in the late 1970s, and by China in the mid-1980s. The former Soviet Union failed to reach this level even by the time of its collapse in 1991. By the 1970s, Japan had more than 2,000 items of annual software acquisition, comprising 65 per cent of its total technology acquisition. From Table 1.3, we can see that the percentage of industrial equipment in the total imports of Japan was consistently low, lower by 12 per cent than the average of developed countries. Most of the imported technology came from the United States at this time (about 65 per cent), which was followed by Switzerland (8 per cent) and West Germany, Britain and France (each 3 per cent) (Ozawa 1985: 229).

Although modern machinery and equipment was imported, large-scale plant imports were relatively low, with the major exception of the power generation industry, which initially depended on the import of large thermal power stations each time generation capacity was upgraded. Even in this industry, the Ministry of International Trade and Industry (MITI) promulgated and implemented a policy by which only one new-type generator should be imported. The others were to be ordered from Japanese heavy electric machinery companies which would purchase the licenses for production. In this way, the three makers of heavy electric machinery, Hitachi, Toshiba and Mitsubishi Electric, received a considerable boost (Hirschmeier and Yui 1981: 310). Similar import substitution was prevalent in other capital goods industries as well.

Historically, Japan has not been very hospitable to foreign investment. The strategy that the Meiji leaders pursued in achieving rapid industrialization was based on their firm determination to maintain both economic and political independence. The Meiji leaders took various measures to restrict the entry of foreign capital, because they perceived direct entry into the Japanese market by foreign firms as being detrimental to the development of Japan's own fledgling industries and technology. After the end of the Second World War, Japan enacted the Foreign Exchange Control Law (1949) and the Foreign Investment Law (1950). These laws were used as the main instrument to control foreign direct investment. Foreign direct investment was permitted in Japan only when it contributed to the attainment of self-sufficiency and the improvement of Japan's balance of payment (Yoshino 1975: 273–290). This restrictive policy was not changed until 1980, when Japan repealed the Foreign Investment Law and thoroughly revised the Foreign Exchange Control Law to liberalize foreign investment and exchange transactions.

Japan's postwar technological progress has been impressive by any standard. In fact, Japan has been widely recognized as the country which has been most successful at borrowing foreign technology (Peck and Tamura 1976: 525–585). The Japanese are seen as care-

Table 1.3 Japanese technology acquisition situation

	1950–54	*1955–59*	*1960–64*	*1965–69*	*1970–74*	*1975–79*	*1980–84*
Annual average items	191	274	825	1,370	2,144	2,013	2,207
Annual average costs (US$100m)	0.1	0.4	1.22	2.56	5.84	9.98	18.68

Source: Adapted from Annual Report of Japan's Technology Import, 1950–85

fully scanning the world for new technology, adroitly controlling the conditions of purchase (or even pressuring reluctant foreign firms to share it with Japanese firms), and concentrating their research efforts on incremental improvements that allow them to outpace the inventors in exploiting the technology. Japan is the only country which was once in a backward position but has now overtaken and in some cases surpassed the most advanced countries in technology.

It is generally accepted that for the late developing countries of contemporary times, real technological independence can be achieved by the software model which reduces dependence on the technology of other countries. But large-scale technology acquisition demands a relatively high technology capability on the receiving side, especially receiving enterprises. The level of software acquisition corresponds to that of the overall technology level. On the other hand, the higher the level of successful software acquisition, the smaller the technological gaps between the receiving country and technologically advanced nations.

There were a number specific conditions for the implementation of Japan's software model. First, Japan had an industrial foundation for seventy years before the Second World War. Although the War wreaked large-scale destruction on its industrial facilities, postwar Japan still kept solid technological and industrial capabilities, including a large number of entrepreneurs and a highly trained and educated labor force. Second, postwar Japan had an acute shortage of foreign exchange and natural resources. It was not possible for Japan to sell primary products in exchange for industrial equipment. Subsequently, Japan had to rely on the software model for technology acquisition. Third, although Japan had a close relationship with the United States and other Western countries, it could not take advantage of the migrant model due to its low level of living standards (in the early postwar period) and very different cultural traditions. Finally, neither the Japanese governments nor private enterprises have had a tradition of actively attracting foreign investment. The lack of interest on the part of Japan's private enterprises in attracting FDI made the large-scale influx of FDI virtually impossible.

CAPITAL MODEL

The model of technology transfer based on the attraction and utilization of FDI is called the capital model. Brazil and Singapore are two typical examples of this model. Brazil is a country endowed with rich natural resources and inhabited by more than 100 million people, whereas Singapore is a small island country, with no natural resources and a population of merely two million. According to a 1985 World Bank report, Brazil and Singapore have followed fairly different paths in attracting FDI. In 1979, FDI in Singapore's GDP was 19 per cent, the highest in the world; while Brazil had 12 per cent, South Korea 3 per cent, India 2 per cent and China (1985) only 1 per cent (World Bank 1988).

Since the end of the Second World War, the capital and technology-intensive industries in Brazil have become very dependent on FDI. In fact, Brazil received almost all of the foreign technology it needed from FDI, which played a crucial role in turning Brazil into a contemporary industrial economy. Multinational corporations (MNCs) provided all the necessary technology and a large portion of capital for Brazil's priority industries. The 1964 coup brought the military to power, who significantly improved relations with Washington and encouraged FDI in the country. From 1965 to 1975, US foreign capital investment in Brazil grew from $1,073 million to $4,784 million. Total foreign capital in the same period increased from $2,861 million to $14,011 million. In the important manufacturing sector, foreign capital grew substantially from 18.9 per cent in 1965 to 28.6 per cent in 1975(Roett 1978: 156–157).

Table 1.4 Initial technology sources of Brazilian enterprises (percentage)

	pre-1930	*1930–45*	*1945–55*	*1956–65*	*post-1965*
Native technology	41.5	52.4	31.7	21.7	28.0
Foreign technology	58.5	47.6	68.3	78.3	72.0

Source: Adapted from UN and Brazilian statistics of the relevant years

Table 1.5 Percentage of foreign enterprises in Brazilian industries

	Fixed assets		*Sales*		*Employees*
	1970	*1977*	*1970*	*1977*	*1977*
Automobile	100	100	100	100	100
Tractor	83	83	80	84	69
Auto part	58	57	63	54	46
Shipbuilding	45	34	30	16	30
Aircraft	36	7	46	20	26
Office equipment	96	91	93	73	69
Electrical appliance	81	86	81	79	83
Industrial machinery	66	51	67	59	54
Precision instrument	40	15	27	24	18
Plastics	73	42	68	57	49
Rubber	67	62	71	81	70
Chemical	54	57	55	57	61
Metallurgical	38	29	36	32	33
Cement	26	41	25	33	17
Textiles	39	37	39	34	26
All industries	34	33	37	44	38

Source: Adapted from UN statistics of the relevant years

What has been the negative impact of the MNCs' activities in Brazil? Peter Evans studied the relationships between MNCs and local firms, their respective bargaining power and the role of the government (Evans 1979). He identified a unique triple alliance in which the government provided infrastructure, local firms the marketing networks, and foreign MNCs the technology and capital. He found that local firms had significant bargaining power in low technology industries, such as the textiles industry, while foreign MNCs had superior bargaining power in high-tech industries. His conclusion is that Brazil has developed with the help of foreign capital, but this is a dependent type of development, where Brazil will remain dependent on foreign MNCs for advanced technology.

A former British colony, Singapore began its self-rule in 1959 and became a republic in 1965. The 1960s witnessed the climax of nationalization in the developing world, but Singapore adopted an exceptional policy toward their development, to attract FDI. From 1970 to 1975 annual FDI averaged US$150 million, from 1976 to 1979 US$380 million, and from 1980 to 1985 US$1.58 billion. By 1982, total accumulated FDI amounted to US$10.5 billion with the main sources being from the United States, Japan, Britain, the Netherlands, etc. The industrial technology of Singapore came mainly from

Table 1.6 The position of foreign companies in Singapore's manufacturing industry

Items	*Numbers*	*Native companies %*	*Foreign companies %*
Numbers of companies (ten or more people)	3,586	75	25
Numbers of employees (10,000)	27.5	45	55
Output value (US$100 million)	170	37	63
Direct export (US$100 million)	102	28	72
Pre-tax profits (US$100 million)	29	30	70
Expenditure (US$100 million)	10	37	63

Source: Adapted from Singapore's Industrial Production Statistic Report of the relevant years

FDI and led to a rise in GNP per capita from US$450 in 1965 to US$4,420 in 1980. FDI helped Singaporeans acquire foreign technology and management techniques and to develop an international market. FDI in Singapore by way of technology transfer was characterized by its quick results and low debt. The success of Singapore's FDI strategy also lay in its advantageous geographical location as a regional trade and distribution center in Southeast Asia (Mirza 1986).

From the recipient's viewpoint, FDI has the advantage of providing a whole set of complicated technology, management and sales techniques and sales channels. In the industries of intensive technology, where rapid change and complicated management processes are involved, FDI can hardly be replaced.

SUMMARY AND CONCLUSION

In this chapter, several generic models of technology acquisition have been discussed and compared. Each model discussed above has its own comparative advantages. Countries adopted their relevant models because of their own respective historical and environmental conditions. It was, for example, impractical for postwar Japan to introduce large numbers of talented immigrants, nor was it possible for the former Soviet Union to absorb large-scale foreign direct investment. Therefore, one has to analyze various historical and environmental factors to understand the best possible national technology acquisition models for specific countries. On the other hand, none of the above countries purely followed one specific model.

While absorbing talented immigrants, the United States has also successfully attracted foreign direct investment. Whilst encouraging foreign direct investment, Singapore has not ignored the acquisition of software and talented immigrants. In fact, most countries in the world have followed a mix of the above models, although the proportion of the mixtures varies a great deal.

It is important for international business managers to understand the possible impact of these different models upon international technology transfer and their investment activities. A country with a clear preference or dislike for one or two models would have national policies providing specific incentives and disincentives. By following the software model, the Japanese government, for example, restricted repeated acquisitions of the same machinery and equipment and discouraged foreign direct investment in Japan. In order to attract foreign investment, Singapore has provided various incentives for multinational corporations to set up their regional headquarters there. Even for those countries having equally mixed models, their trade and investment policies in terms of technology acquisition can be better appreciated with these models in mind.

2 Assessing Japanese technology transfer to Southeast Asia

INTRODUCTION

Nowhere in the world is the influence of transferred Japanese technology greater than in member nations of the Association of South East Asian Nations (ASEAN). Being an earlier modernized member of Asia and geographically proximate to ASEAN member nations, Japan has been regarded by the Southeast Asians as the natural source of technology for the region, as is evidenced in the Look East Policy of Malaysia beginning in 1981. For Japan, Southeast Asia has always been a crucially important source of raw materials and cheap labor for its dynamic manufacturing industry. Furthermore, Southeast Asia has become not only a springboard for Japanese products to West European and North American markets, but also itself one of the fastest growing markets for Japanese products. Japan's technology transfer via FDI to Southeast Asian countries has been designed to strengthen and develop its ties with the region.

Compared to most Western advanced nations, Japan is a relatively new actor in the area of technology transfer. For the first three decades of the postwar period Japan was more a recipient than a source of transferable technology, though the first postwar export of technology to the developing nations of Asia can be traced back to the early 1950s. The 1960s witnessed a dramatic increase in the export of Japanese technology to Asia and beyond. Except for a short period in which expansion was slowed down by the oil crisis in the early 1970s, this dynamic growth trend continued in the following two decades and the technology transfer has been diversified. This chapter examines the major channels and determinants of this transfer, and analyzes the advantages and problems in the overall technology transfer of Japanese companies.

MAJOR CHANNELS OF JAPAN'S TECHNOLOGY TRANSFER TO SOUTHEAST ASIA

The nature of the technology that Japan transfers to advanced industrialized countries is fundamentally different from that of the technology transferred to the developing countries of Southeast Asia. Technology transferred to the advanced industrialized countries largely consists of patented high-level technology while that transferred to the developing countries is mainly modernization experience and skills closely related to standardized production methods. The scope of a typical technology transfer contract usually covers production, management and marketing. The various production activities that Japanese companies have transferred to Southeast Asia include: material selection, selection and installation of equipment, plant layout, assembly methods, machine operation, training of personnel, maintenance techniques, provision of technical data, quality and cost controls, and inventory management (Ozawa 1981: 3–6).

Direct foreign investment constitutes the key channel through which most of Japan's technology is transferred to the member nations of ASEAN. The unusual concentration on direct investment forms a sharp contrast to the numerous licensing agreements Japanese companies have signed in transferring technology to companies in advanced industrialized nations in the West. In the 1980s Japanese FDI in developing countries caught up with and surpassed that from the United States. By the late 1980s, Indonesia and Singapore continued to be the chief beneficiaries of Japanese investment in ASEAN, with an average share of 55.9 per cent and 14.4 per cent, respectively, followed by Malaysia with 10.9 per cent, Thailand with 9.7 per cent, and the Philippines with 7.7 per cent (Chew *et al.* 1992: 115–116).

Until the mid-1980s, joint venture seemed to be the preferred means for Japanese companies to enter host countries in Southeast Asia. Policies of the Southeast Asian countries to restrict imports while promoting local production also encouraged foreign direct investment. Since there were various rules to restrict the shareholding of foreign investment before the mid-1980s (like the 1974 rule of Indonesia that all foreign ventures must eventually be transformed into minority holdings), many Japanese companies felt a necessity to form joint ventures. There was also a widely shared feeling among Japanese managers that joint ventures were not as vulnerable to political risk and local nationalism as wholly owned investment.

One notable motive for many Japanese companies to select direct investment as a transfer vehicle was derived from the nature of the transferred technology. For a long time, Japanese companies almost exclusively transferred general know-how and industrial experience. The transfer of this type of technology entails long-term involvement by the transferor in the production and management activities in the host country. Moreover, technology recipients tend to require foreign investors to be involved in the initial stage of production. Many developing countries do not usually recognize the economic value of industrial expertise and tend to regard it as a free service that should accompany the purchase of machinery and equipment. Therefore Japanese companies found it necessary to obtain sufficient compensation for their technology through capital ownership and direct management of their foreign investment (Ozawa 1981: 40).

Traditionally, Japan's direct investment in Southeast Asia was fairly small in scale compared to that coming from the United States and Western Europe. Japanese investment in the region tended to be concentrated in relatively small shares of capital ownership and less capital-intensive operations. The overall pattern was influenced by host country policies to encourage local participation in ownership and the Japanese tendency to invest in labor-intensive industries. Such a tendency reflected the desire of many investing Japanese companies to take advantage of cheap Southeast Asian labor (Ozawa 1981: 8–10).

Nevertheless, this situation changed drastically since the Plaza Agreement in 1985 as Japanese FDI more than tripled between 1985 and 1988 (US$12.2 billion to US$47 billion). Developed countries have attracted the majority of investment, approximately 60 per cent up to the end of the fiscal year 1988 (March 1989). The developing countries in Asia have received a large share, with 45 per cent of total FDI approvals in developing countries in the fiscal year 1988 going to the Asian developing economies. Asia has also accounted for 64.1 per cent of Japanese manufacturing FDI in developing countries (Urata 1991: 175, 179). According to the *Tokyo Keizai Shinposha*'s survey, the number of Japanese-affiliated companies operating in Asian countries at the end of 1988 was 3,770, with ASEAN countries accounting for 1,805. If indirect employment (through subcontract) is counted, Japanese investment accounted for the employment of three million workers (Yamashita 1991: 9–10).

KEY DETERMINANTS OF JAPAN'S TECHNOLOGY TRANSFER

The Japan Islands are relatively poor in mineral resources, with the exception of sulphur and pyrites, which are byproducts of frequent volcanic activities. Although Japan still has coal deposits in Hokkaido and Kyushu, there is little good coking coal which is used in steel making. Foreign sources provide 84 per cent of its iron ore, 89 per cent of its coking coal and also 100 per cent of the nation's petroleum, aluminum and uranium (MCA 1993: 350–351). The lack of geological resources constitutes a key reason behind Japan's technology transfer to Southeast Asia. Japan depends on overseas supply for all major industrial raw materials, much of which come from there. In fact, the one key reason for the Japanese military to launch the Pacific War in 1941 was to secure the supply of oil and other vital materials from Southeast Asia. The postwar economic boom and rapid development in manufacturing industries have resulted in a dramatic increase in Japan's need for critical resources.

Since the early 1960s, Japanese demand for raw materials grew to account for a substantial share of world consumption. This development pushed Japanese resource industries to increase investment in exploration and development of natural resources overseas. A significant portion of earlier Japanese investment in Southeast Asia was concentrated in developing natural resources for the domestic consumption of Japanese industries. Examples of such FDI included petroleum drilling in Indonesia, iron ore mining in Malaysia and copper mining in the Philippines. Subsequently, ventures involved in natural resources development played a dominant role in transferring technology during this period.

The major players in developing natural resources in Southeast Asia were the "Big Nine" general trading companies (*sogo shosha*), which include Mitsubishi, Mitsui, Marubeni-Iida, Itochu, Nissho-Iwai, Tomen, Sumitomo, Nichimen and Kanematsu. The worldwide sales and purchasing networks of these general trading companies were crucial to Japan's trade with Southeast Asia and the world. Traditionally, the role of the trading companies was to assist Japanese manufacturers in importing raw materials and exporting their products. As a consequence, they became the pioneers of Japan's technology transfer to developing countries (Yonekawa 1990). More recently these companies have begun to make investments in the development of natural resources located overseas. By the early 1980s, the general trading companies' overseas investment constituted one-third of Japan's overseas investment (Yonekawa 1990).

From the 1960s until most recently, Japan experienced an acute labor shortage. As a result, there was constant pressure to raise wages to compete for the scarce labor force. Unlike other advanced industrialized nations in the West, Japan did import workers from developing countries with abundant cheap labor. Foreign laborers from Southeast Asia and Brazil (mostly ethnic Japanese) did find ways to get into Japan although they constituted less than 2 per cent of the total labor force and were mainly concentrated in jobs in which few Japanese workers would be interested. As an alternative to the labor shortage and rocketing labor costs, many Japanese companies have chosen to export their manufacturing facilities and technology to nations of Southeast Asia, where they can take advantage of cheap labor to maintain their competitiveness in the world market.

In the 1980s, a combination of changes prompted many Japanese companies to greatly increase their investment activities in ASEAN countries. First, the yen appreciation caused the price competition of Japanese manufactured goods to decline in the world market. Japanese companies had more incentives to shift their production operations to countries with low wages and production costs. Secondly, in order to diffuse trade frictions with the United States and Western European nations, Japanese manufacturing companies had to diversify their production and distribution systems. One effective strategy has been to construct factories in ASEAN countries and distribute their products to the United States, European Union member nations and other countries directly from those locations. Finally, in most Southeast Asian countries, government policy was drastically changed in the mid-1980s from one emphasizing import substitution industrialization to one of export orientation. ASEAN governments also accompanied the deregulation measures with invitations for more foreign investment and with a vigorous promotion of exports (Phongpaichit 1991: 39–40).

According to an investigation of the ASEAN Promotion Center on Trade in Tokyo in 1989, of the companies which invested in the ASEAN region, those citing the use of low-cost labor as their main reason accounted for 61.2 per cent, higher than in any other area. The second most commonly quoted reason was the cultivation of the local market, cited by 40.1 per cent. Among the companies planning to invest or interested in investing in the ASEAN region, the use of low-cost labor was also the main reason (58.8 per cent). Nevertheless, export to Japan or a third country ranked second (36.9 per cent), followed by cultivation of the local market (33.6 per cent) and the securing of inexpensive raw materials (27.2 per cent) (Tokunaga 1992: 17).

THE ROLE OF JAPANESE GOVERNMENT

The Japanese government has played a proactive role in promoting technology transfer. It is not only directly involved in providing technical assistance to developing countries, but also offers various incentives to private companies engaged in technology transfer. Among them are three governmental measures to facilitate technology transfer: technology export financing, export insurance and tax credits on income from technology exports.

Japanese overseas technical cooperation was formally launched in 1962 with the establishment of the Overseas Technical Cooperation Agency. In 1974 this agency was merged with the Japan Emigration Service and the Japan Overseas Agricultural Development Foundation to form the Japan International Cooperation Agency (JICA). The main functions of JICA include:

1 Government-sponsored technical cooperation.
2 Grant-aid cooperation promotion programs.
3 Dispatch of Japan Overseas Cooperation Volunteers (JOCVs).
4 The Development Cooperation Program for investment in and financing of development projects.
5 Recruitment and training of qualified Japanese experts for technical cooperation.
6 Emigration services.

In 1987, JICA expended 62.8 per cent of the Japanese government's total budget for technical cooperation programs. As of 1987, the major portion of this money went to Asia, and the biggest share of that to Southeast Asia. Technical cooperation projects under JICA range from light and heavy industries, chemical industry, transportation, and construction, to agriculture, fisheries, public administration and management techniques. In 1987, the largest number of trainees came from Thailand (8.4 per cent), followed closely by Indonesia (8.2 per cent) and Malaysia (7.5 per cent). The largest number of Japanese experts were dispatched to Indonesia (12.7 per cent), followed by Thailand (11.4 per cent). A substantial number of ASEAN trainees came from the human resources area. Other significant areas included manufacturing and public health and medicine (Thailand), agriculture (Malaysia), agriculture and public administration (the Philippines), and commerce and trade and social infrastructure (Singapore) (Chew *et al.* 1992: 120–121).

Technology export financing constitutes a very important factor for the success of technology transfer. Under the amended Export-Import Bank Act of 1986, the Export-Import Bank provides Japanese companies with overseas investment credits to fund their overseas investment as follows:

1 Credits to Japanese companies for equity participation in foreign companies.
2 Credits to Japanese companies for loans to foreign governments or companies to provide them with long-term funds for ventures operating outside Japan.
3 Credits to Japanese companies to be extended as loans to foreign governments or companies for their equity participation in foreign companies in which the Japanese companies have equity shares.
4 Credits to Japanese companies for their equity participation in companies established in Japan for the sole purpose of making overseas investment in the three above-mentioned items.
5 Credits to Japanese companies to provide them with funds required for projects operating abroad.

(EXIM 1989)

The Export-Import Bank of Japan extends loans to companies at relatively low interest rates. The interest rate is based upon such factors as the type of investment project and current market interest rates. Repayment terms in most cases are from seven to fifteen years, depending on the profitability and cash flow of the project. EXIM also extends buyer credits to foreign companies or financial institutions in order to finance the purchase of Japanese goods and services. Buyer credits cover up to 85 per cent of the export contracts. By the late 1980s, the EXIM Bank mainly operated foreign investment schemes, i.e., overseas credits and overseas project loans, rather than original official export credits. For example, the percentage of overseas investment credits to financing disbursements rose to 37 per cent in 1990 from 25 per cent in 1985, while the share of export credits declined to 12 per cent from 51 per cent during the same period (OECF 1990).

Export insurance constitutes the third important factor in promoting technology transfer. Before 1987, the Export Insurance Act of 1950 was widely used in technology exports. Exporting companies were protected from default by an importer in return for an insurance premium paid to the Ministry of International Trade and Industry

(MITI). In 1987, the Export Insurance Act was amended and renamed the Trade Insurance Act, under which MITI has established a new trade insurance system. Under the new system, MITI provides protection against risks or losses from overseas investment, intermediary trade outside Japan, and prepaid imports as well as exports. For example, before 1987, Overseas Investment Insurance was provided to cover only losses due to events with the nature of political risk like war or internal disorder. The new and expanded Overseas Investment Insurance insures against losses caused by commercial risks, including both bankruptcy linked with FDI and political risks. The new Overseas Investment Insurance has a clear goal of promoting the establishment of production facilities abroad by small and medium-sized Japanese companies (Tokunaga 1992: 23–24).

Finally, tax credits on income from technology exports have also been conducive to technology transfer. The Japanese have a beneficial system whereby 70 per cent of corporate profits earned from technology exports may be excluded from taxable income and MITI grants the right to use this tax shield (Ozawa 1981: 42). In addition, an indirect foreign tax credit allows companies to treat investment abroad through a foreign subsidiary in the same way as investment through a branch. Foreign taxes that are levied on the foreign subsidiary may be credited against the Japanese tax levied on the Japanese parent corporation. The government also allows a 5.9 per cent special deduction for income derived from overseas technical service transactions (Ishi 1989: 181).

To facilitate technology transfer to ASEAN, the Japan ASEAN Investment Company (JAIC), in which the OECF has equity investment, was established in 1981 with 137 member corporations of the Japan Association of Corporate Executives (Keizai Doyukai) as shareholders. The JAIC-1 Investment Fund, amounting to Y7 billion, was established in 1988 and lasts for ten years from the date of establishment (Tokunaga 1992:24). As of January 1991, investments had been made in fifty ASEAN corporations. The assets are jointly owned by its members based on the ratio of their initial investment in the Fund.

In addition to the direct involvement of the Japanese government in promoting technology transfer, it also sponsors or supports various organizations promoting the export of Japanese technology. The Japan Plant Export Association was established in 1955 and the Japan External Trade Organization (JETRO) was founded in 1958. For the past few decades, JETRO has vigorously pushed for export through its trade shows, overseas market research and advertisement.

ON-THE-JOB TRAINING IN JAPAN'S TECHNOLOGY TRANSFER

As most technology transferred by Japanese companies to Southeast Asia is related to labor-intensive industries, labor training occupies a prominent position in the Japanese strategy of technology transfer. For this reason, on-the-job training (OJT) has been considered by some as Japan's "inner mechanism of technology transfer" (Ozawa 1981: 35). OJT not only provides technical and administrative knowledge to the workers, but also teaches them how to have higher motivation and better discipline so that the process of never-ending quality improvement can be fulfilled. Unlike European and American companies which basically utilize written manuals and detailed job descriptions, Japanese affiliated manufacturing companies base their production management methods and their technical training on OJT.

Japanese affiliated companies have different approaches toward technology transfer. Most European and American companies will pull back their technical advisers when the factory runs smoothly; and the local employees will only need to follow manuals carefully. But in Japanese affiliated companies, technical advisers tend to stay even after a good operation has been achieved. They will continue to train the workers step-by-step in productivity and quality control, maintenance and repair, utilization of new production methods and new technology, as well as other production-related skills (Yamashita 1991: 18).

There are a number of reasons for the Japanese to adopt such an approach (Hieneman *et al.* 1985: 145). First and foremost, the technologies transmitted via OJT are basically know-how or experience related to well-proven and standard production techniques. Technology of this type cannot easily be transferred either in the form of industrial equipment or through blueprints or operating manuals. Instead, it can be better transferred through personal communication between workers and managers at all levels. Moreover, for most Japanese expatriates the language barrier poses a particular difficulty in communication, as most of them do not have a sufficient training in local languages and their constant job rotation makes language learning ever more difficult. This problem may help explain why Japanese managers tend to like the "learning by doing" approach in transferring technology rather than depending on comprehensive manuals that a large number of under-educated workers may have trouble understanding.

By adopting OJT, Japanese affiliated companies also hope to

continually improve technology at the shop floor level (Kimbara 1991: 163–164). As technology continually progresses to a higher level, it can hardly be written into the manual thoroughly. For the Japanese, there should be no end to technology improvement. Technological progress is considered a dynamic and incremental process, and must be pursued by all members of the organization rather than only by engineers. Therefore, Japanese workers on the shop floor are also involved in the activity of technological improvement. This conception is clearly manifested in the quality control that symbolizes the unique strength of Japanese production management. The Japanese excel in continuing to improve the quality of their products, the process commonly called *kaizen*. The effect of such incremental innovation is highly visible when the product or technology is standardized.

The heavy reliance of Japanese companies on "transfer through people" is also closely related to their emphasis on direct foreign investment as a major channel of transfer. There is usually a strong linkage between a supplier company's willingness to be involved in the training of the local employees and its financial stake in the recipient. In the case of licensing and technical cooperative arrangement, training programs are much less significant. For example, the five Asian countries that accounted for over half of the total number of trainees sent to Japan (Thailand, Republic of Korea, Taiwan, India and Malaysia) during the 1970s, were also the largest recipients of Japan's direct investments in manufacturing (Ozawa 1981: 36).

As has already been discussed in the previous section, the Japanese government is committed to a policy of encouraging and supporting technical training for these developing countries. Large numbers of trainees have been sent to Japan for technical instruction under various programs sponsored by the Japanese International Cooperation Agency (JICA).

MAJOR PROBLEMS IN JAPAN'S TECHNOLOGY TRANSFER

In spite of relatively conspicuous success in technology transfer via OJT, the transfer of technology from Japan to Southeast Asia has not been problem-free. The sheer size of Japanese investment in the region itself may constitute a major source of trouble. Anti-Japanese sentiment is deep-rooted and evident among Southeast Asian people, who fear that Japanese capital may become another version of Japan's dominance in Southeast Asia. Concerns have been expressed on various occasions about the over-presence and over-influence of Japanese capital. Like other foreign capital, Japanese capital has also worsened the gap of income distribution of host countries. Most partners of Japanese joint ventures in the ASEAN countries, for example, are wealthy overseas Chinese, who may obtain substantial concessions and profits by having successful joint ventures with the Japanese.

Cultural and business styles have also posed problems for the Japanese. The use of traditional Japanese management techniques in the ASEAN member nations has tended to create difficulties. Lifelong employment, seniority-based promotion, the *sempai-kohai* (mentor) system, and consensus decision-making serve to promote strong cooperation between management and workers and generate lasting company loyalty in Japanese companies in Japan. Their famous quality control and improvement process have enabled Japanese manufacturers to produce high-quality products. Nevertheless, the Japanese have encountered serious problems in grafting their management system to the local environment.

Localization of managerial resources has developed, but local people in ASEAN member countries are critical of Japanese affiliated companies when compared with their Western counterparts. This is in accord with Ah Ba Sim, who conducted research on the organizational relationship between Malaysian subsidiaries and their foreign parent companies. There is much less decentralization in the Japanese subsidiaries compared to their American and British counterparts. There are, for example, few (and sometimes no) expatriates in Western subsidiaries in Malaysia, while in Japanese companies Japanese expatriates often account for about 1 per cent of the total number of employees (Sim 1978). The ratio of Japanese at the section chief level stands at around 44 per cent (Kawabe 1991: 265). Subsequently, local managers feel that they are often denied sufficient responsibility and control. For most of those local managers, the prospect of being promoted to the highest level of management seems to be very remote.

Traditionally strong group identification also serves to alienate Japanese expatriates from their employees and causes conflicts in management and technology transfer to Southeast Asia. For example, when locals organize tennis games and dancing parties, Japanese managers often leave quickly after giving a speech, while this kind of practice is not found among Western managers (Kawabe 1991: 259). The language barrier further worsens the problems of communication. Even if a Japanese manager makes a special effort to overcome the language and cultural barriers in Southeast Asia, he will usually be rotated out of his position in two to three years. This management practice of Japanese companies serves to dampen any desire among Japanese expatriates to mix with locals.

The manner in which most Japanese companies handle technology transfer has also often been criticized by locals as reflecting the Japanese unwillingness to teach more sophisticated technology to the local people. Japanese managers have tended to show insufficient confidence in local employees and consider it more appropriate to design and develop new products at the headquarters research centers in Japan. Japanese affiliated companies tend to transfer technology that is necessary mainly for routine operations. In Malaysia, for example, operation and maintenance technology are transferred by 75 per cent and 54.2 per cent of Japanese affiliated companies respectively. In Thailand, operation and maintenance technologies are transferred by 78.7 per cent and 57.4 per cent respectively, but other kinds of technology are transferred in less than half of the Japanese affiliated companies. Furthermore, the heavy reliance on OJT or on the Japanese technician's experience sometimes causes serious misunderstandings between workers and managers (Kimbara 1991: 163).

Even in OJT, Japanese affiliated companies have encountered a serious problem which is mainly the relatively high rate of turnover of the trainees once they return to their respective companies in their home countries. Lifetime employment is not part of indigenous traditions; and the commitment of workers to their companies is much less than that of the Japanese. When skilled workers return home, they are usually in high demand in the job market and find it hard to reject more lucrative offers from other companies. The famous quality control process that OJT is aimed at achieving has proved to be more difficult to transfer. In some local cultures, such as in Indonesia, manual labor is perceived as low status and this may result in a lack of attention to physical detail or quality control (Price Waterhouse 1989: 24).

Since the beginning of the 1980s, the Japanese have made increasing efforts to deal with the problems accompanying their technology transfer and direct investment to Southeast Asia. More and more of them have begun to examine the applicability of the Japanese management system and the possibility of a higher degree of localization and decentralization. With the further diversification of production from purely labor-intensive industries to more complicated manufacturing processes, pressure has built up to expedite higher-level technology transfer. As an increasing number of ASEAN trainees return home, Japanese companies have begun to hire them for higher positions of management.

In Malaysia, for example, Japanese subsidiaries have recently begun to hire Malaysian students who studied in Japan through the auspices of two programs that started under the Malaysian Look East Policy. One program is the dispatch of Eastern Technical Trainees, who were selected from government organizations, business companies and educational training organizations to receive training based on OJT in the corresponding Japanese organizations. The second program sends Malaysian students to Japanese universities, where they are usually enrolled in such practical subjects as engineering, management and economics. Many of these returned students were hired by Japanese subsidiaries in Malaysia. The problems caused by the dual structure of Japanese affiliated companies in the process of globalization may be mitigated by the employment of such students, because the role of Japanese expatriates in translating Japanese systems into local ones may be more readily accepted by the students. The eventual creation of a hybrid culture in the Japanese affiliated companies may be achieved (Kawabe 1991: 265–266).

SUMMARY AND CONCLUSION

This chapter first discussed major channels of Japan's technology transfer to Southeast Asia, then the key determinants of Japanese technology transfer were explained. This was followed by analyses of the proactive role played by the Japanese government and the on-the-job training management in Japanese technology transfer. Existing problems are examined at the end of the chapter.

Japan is currently the main source of technology for Southeast Asia. The Japanese have proved to be very successful in packaging mature technology for transfer to the developing countries of Asia. Industrial expertise and know-how have been the primary transfer while direct foreign investment constitutes the most widely used transfer channel. Traditional Japanese OJT management has commonly been used to assure success of the transfer process. Japanese governmental support has facilitated the endeavor of Japanese companies to transfer their technology to the region. The continued need for a secure resource base, a cheap labor pool and the cultivation of local markets will continue the flow of technology and investment from Japan.

Japan has a special comparative advantage in technology transfer.

The Japanese have had successful experiences in adapting foreign technology to their own industries and providing their workers with sufficient training in utilizing the transferred technology. Therefore, the Japanese are in a unique position to transfer technology to the developing countries of Southeast Asia. The process of transfer is also beneficial for the Japanese, since Japan needs to shed some of its traditional industries in order to promote high-technology and service-based industries. On the other hand, the continued flow of technology and investment from Japan to the ASEAN member nations will not only contribute to their rapid industrialization but also help integrate the economies in the booming Pacific Rim.

3 Comparative host governments' FDI policies

INTRODUCTION

Foreign direct investment (FDI) has made major contributions to world development in the past forty years. Capital flows have been very beneficial to countries with poor domestic sources and limited capability to raise funds in the world's capital markets. The role of FDI has been seen as that of technology transfer (Casson 1979: 1). Since the end of the Second World War, ideology toward FDI has ranged from the non-interventionist principle of free market economy to a radical anti-FDI approach. In between these two extremes is what might be called pragmatic nationalism. Policies based on these three approaches can be somewhat indistinctly categorized as "open," "restrictive," or "mixed." But very few economies can be called completely open or closed; and various national policies are along the continuum between the extremes (Hill 1994: 197; Behrman and Grosse 1990: 107).

The policies toward FDI can also be analyzed in the context of the basic development strategy of each country. Since the end of the Second World War, three different strategies of economic development have been practiced: autarchy, import-substitution and export-driven (Todaro 1989: 427–465). Until 1980, autarchy was the principal strategy for many socialist and newly independent nations which basically rejected FDI. During the 1950s and 1960s, many developing countries, especially in Latin America, followed an import-substitution strategy. While FDI inflows were used to develop selected native industries, they were subject to various restrictions. Export-driven strategies have been practiced by countries with relatively small internal markets. Because multinational corporations (MNCs) can provide capital and market access, as well as technology and managerial skills, FDI inflows have been encouraged.

This chapter compares the three categories of national policies toward foreign direct investment.

OPEN POLICIES

The selection of an open policy by a government has nothing to do with the level of development, but rather reflects its view on how the economy can best be developed. Countries that are most open to foreign MNCs' activities are those that uphold a free market view, which has its roots in classical economics and the international trade theories of Adam Smith and David Ricardo. The free market view argues that international production should be distributed among nations based on the theory of comparative advantage. The MNC is regarded as an instrument for dispersing the production of goods and services to those locations around the world where they can be produced most efficiently. FDI by the MNC is a way to increase the overall efficiency of the global economy (Reich 1991).

Open countries impose few limitations on MNC activities and provide national treatment to foreign firms. Most of these countries are "home" to MNCs' parent companies. The United States and Canada represent the least restrictive nations toward MNCs, followed by Germany and the United Kingdom. Among newly industrialized countries (NICs), two of the Four Tigers of Asia (Hong Kong and Singapore) have very few limitations on foreign business entry, though they provide guidance to enterprise in general. A few less developed countries, notably Chile, have developed highly market-oriented regimes to attract more foreign capital and technology to their economies. In addition, several countries have created unrestricted legal environments to attract specific kinds of business such as banking and finance (e.g. Panama, the Bahamas, the Cayman Islands and the British Channel Islands). Nevertheless, even the most open economies keep various controls over some sectors for the protection of national security or public interests.

To illustrate open policies, we can consider the US policies toward FDI. The United States has maintained a "free-market" system as the mainstay of its economic development. Since 1933, it has advocated the principles of freedom of trade and capital movement and has consistently opposed protectionism in foreign trade and investment, and the imposition of direct controls. Generally, it has been more liberal in capital movement than in trade. Presently, the United States places few restrictions on foreign direct investment. US policies on foreign firms are identical to those on domestic American firms (Jackson 1991).

Some exceptions to this openness do exist, such as the restrictions on firms that supply products and services to defense-related industries. More specifically, some activities in the areas of designing, producing and installing military equipment are limited to domestic firms, although the components used in such equipment may be imported or made by foreign-owned firms in the United States. Furthermore, the United States does not allow the purchase of US businesses by agents of several Communist countries (such as Cuba, North Korea and Vietnam). Other Communist government agencies are permitted to operate in American markets, especially in international banking and trading activities. Restrictions are also imposed on firms from other countries deemed unfriendly to the United States, such as Iran and Libya.

The US government also owns a few businesses. The government-owned postal service has been granted monopoly rights to provide first-class mail delivery. But the scope of this business has been limited, as the government allows others to provide express mail, regular package delivery and other forms of document transmittal, such as telex and fax. In contrast, the government Post-Telephone-Telegraph (PTT) in many other countries maintains a monopoly over postal and telecommunications services. Other business participation includes ownership and management of a major electric power utility, the Tennessee Valley Authority, and of Amtrak passenger rail service in the Northeast. The US government also reserves some business activities for domestic providers, such as the press, radio and television transmission, freshwater shipping and domestic air transportation. Each of these industries is limited to firms owned at least 75 per cent by US interests. Finally, work on classified US government contracts, development of federally owned lands and operation of nuclear power facilities is restricted to US firms.

There are no federal limits on foreign firms buying US real estate, financial institutions, industrial companies or other direct or portfolio investments. Some state and local governments in the United States do have laws restricting the activities of foreign nationals. New Hampshire, for instance, allows mineral prospecting and mining only by American citizens. However, their ability to regulate foreign ownership is limited by the constitutional prohibition on restriction of interstate commerce (Graham and Krugman 1991: 133). Therefore, any foreign firm from a friendly country can start a business in the

United States by following the incorporation steps for a similar domestic company.

In spite of its open policy, the increasing inflow of FDI and the rising numbers of acquisitions and mergers from abroad has caused growing concern in the country. The Committee on Foreign Investment in the United States (CFIUS) was established in 1975 within the executive branch to review FDI inflows that might adversely affect US interests. CFIUS monitors investment trends, provides guidance on prospective inflows, and prepares proposals for new legislation or regulations on FDI, but it does not have authority to approve or disapprove FDI inflows. In 1988, Congress added a provision to the 1988 Omnibus Trade Bill restricting FDI inflows that were deemed to adversely affect "national security." This provision, known as the Exon-Florio Provision, expands the authority held through some other avenues to restrict FDI inflows (Graham and Krugman 1991).

The Exon-Florio Provision was made due to the proposed purchase by Fujitsu (Japan) of the Fairchild Semi-Conductor Corporation, which was previously bought by the French company, Schlumberger. Because the Emergency Powers Act requires declaration of a national emergency regarding the relevant foreign country, it could not be applied to this case. Many in Congress believed that the President should have wider authority to block a foreign takeover when necessary. This new provision gave the President the authority to block or suspend acquisitions, mergers or takeovers by a foreign person or entity that he considers might threaten national security. The administration did not challenge the provision as improper or against the generalized "open policy." Under this provision, a review and presidential decision are to be completed within ninety days. The process of review and recommendation is delegated to CFIUS, while the Department of Treasury implements the regulations.

In many aspects, the policies of the open countries are similar to those of the United States. Except for reserving some public services such as telecommunications, media and air transportation for local entrepreneurs or the host government itself, open countries do not restrict the entry or exit of FDI. Tariffs are very low in most of these countries, and other import restrictions such as licensing requirements are not normally imposed. International financial flows are generally unrestricted. Price controls, local content requirements and other competitive restrictions are rarely used. Most importantly, they grant foreign firms national treatment. In sum, open countries are more receptive of and also more attractive to MNCs than those with restrictive and mixed policies.

RESTRICTIVE POLICIES

Restrictive policies are closely related to the radical views that trace back to Marxist political and economic theory. Radical theorists argue that the MNC is an instrument of foreign imperialist domination. They see the MNC as a tool for exploiting host countries and exclusively benefiting the capitalist home country. According to the radical view, FDI by the MNCs of advanced capitalist countries is used to keep the underdeveloped nations in a dependent position on advanced capitalist countries for capital, technology and jobs. An extreme version of this argument maintains that no nation under any circumstance should ever allow the inflow of foreign direct investment. Moreover, if MNCs already exist in a country, they should be nationalized (Hood and Young 1979).

The countries following restrictive policies are those carrying out an autarchic economic development strategy. But very few countries in the world can be strictly put into this category. The most typical restrictive countries include Iran, North Korea and Burma. Naturally, none of them is a major destination of foreign direct investment. Historically, China was an extreme version of this category, as the then Communist leader Mao Zedong was an adamant follower of a self-reliance strategy and closed China not only to Western countries but also to most socialist countries. Until recently, India was a marginally restrictive country.

Generally, these nations have tried to severely restrict the presence of foreign firms in their economies, if not totally prohibit them. The economic policies of many socialist nations not only limit the ownership of foreign companies, but also control their activities in various ways. In some countries, restrictions on access to foreign exchange place tremendous stress on foreign business, since imports are often needed and profit remittance is subject to governmental control. Many nations have followed a so-called "zero-balance" policy on foreign exchange, which requires foreign-owned firms to generate at least as much foreign exchange through exports as they need to import production equipment and components and remit profits. In addition to limits on ownership and financial transfers, different countries have used a wide range of rules and regulations to constrain the activities of MNCs.

The main reason for using India as an example of this group is that people tend to think that only Communist and socialist countries have followed restrictive policies. The largest democracy in the world with a basically market-oriented economy, India had for four decades adhered to a socialist ideology and applied restrictive policies toward foreign firms. Particularly since 1973, India's policies had been inward-looking and socialist, even though India did not have the central planning of socialist countries. Foreign firms interested in the Indian market were limited to export–import contracts, cooperative arrangements and minority joint ventures with Indian partners.

Since independence from Britain in 1947, India had been reluctant to allow participation of foreign firms in the economy. Both Nehru and Indira Gandhi attached great importance to self-reliance for India. The Foreign Exchange Regulation of 1973 imposed stringent controls on foreign exchange and also put foreign firms under the regulation of the Reserve Bank of India. Foreign ownership was restricted: 100 per cent ownership could be retained only for those exporting all of their production or using high technology; some high-tech industries, like computers and optical fibers, allowed foreign companies up to 75 per cent ownership; foreign companies could own up to 51 per cent if other desired technologies were used (such as turbines) and a prescribed portion of products was destined for export; and most others were only entitled to 40 per cent or less ownership, thus essentially becoming Indian firms (Behrman and Grosse 1990: 184).

There has been a limit on FDI in sectors set aside for the state-owned enterprises (SOE) or to the local private sector. The industries reserved for SOEs included: oil, iron and steel, most mining and production of mining equipment, aircraft, shipbuilding, arms and ammunition, heavy electrical equipment, all utilities, and others. Beyond this, FDI was generally not permitted when the desired technology and products in a sector were available locally. In fact, approval of FDI was granted mainly to those industries designated as "core sectors" in which technology or exports were highly sought after. In all other industrial sectors, investment was limited to locally owned companies.

For the industries open to FDI, the process of obtaining approvals was very complicated: one basic license and three or more official endorsements were to be received. First, the foreign firm had to obtain an industrial license under the Industries Act of 1951. For a small foreign firm, or one that planned to export a large part of its Indian production, this license was not very hard to get. For most other firms, the process could take a very long time and be very complicated. The second key permission was an approval by the Foreign Investment Board concerning the terms of the FDI and especially the arrangements with the Indian partner. Third, permission under the 1947

Capital Issues Act had to be given to issue new capital stock. Finally, the foreign firm had to get a license to import capital equipment related to the project. The time needed to fulfill all of these requirements was normally no less than four months, and it often took almost two years (BIC 1987: 6).

It is not surprising that the FDI level in India has been very low. Many MNCs which were already in India, like Coca-Cola and IBM, had to find their way out of India. In the period from 1969 to 1982, India approved fresh equity inflows of merely $80 million in total. Of these approvals, less than half materialized, so that only about $40 million was actually invested in the country during the fourteen-year period. This formed a sharp contrast to a net inflow of $14 billion in 1970–80 to Brazil, $7 billion to Mexico and $1.5 billion to Argentina which had a stagnant economy (Lall 1985: 49).

Until the mid-1980s, the Indian government held the position that the enormous domestic market presented such an attractive prospect that foreign firms had to accept unpleasant conditions in order to gain access to it. Nevertheless, the government did grant exceptions to the rules for foreign companies that had at least either key proprietary technology or access to foreign markets for exports. Some of these companies were permitted up to 74 per cent equity ownership in their affiliates (Behrman and Grosse 1990: 186).

In addition to local ownership requirements, the Indian government also tried to maximize local content in production. Import licenses for production input were normally issued only for products that could be manufactured locally. Generally, import restrictions were enforced through various mandatory licenses for all imports. Until the late 1980s, India maintained very high tariffs on all products. These numerous rules and regulations were also very complex. Tariff rates ranged from 30 per cent to 100 per cent of the value of the shipment. In short, India's import policy had been in conformity with its overall restrictiveness toward foreign business. These trade restrictions were also designed to encourage technology licensing and investment projects in priority industries.

Generally, limitations on MNC activities in restrictive countries are extremely harsh in comparison with those in the open category. In restrictive countries, FDI entry is approved only under highly regulated conditions, and even then with minority MNC ownership usually required. Wide-range price controls are exercised throughout the group. Imports are subject to severe restrictions, with some products not excluded and others requiring licenses or some form of government approvals. In many cases, imports are allowed only through a counter-trade deal. Financial flows are highly restricted, because most of these currencies are inconvertible. Although there are many differences in the treatment of foreign MNCs among the countries in this category, all of these countries try to keep their business activities under the control of local people. In short, their business policies are highly interventionist.

Since the beginning of the 1980s, the number of the countries which adhere to radical ideology and a restrictive approach has been diminishing. Most notably, China opened its doors in 1979 and has since been actively seeking foreign direct investment (Kleinberg 1990). In fact, China became the second hottest destination for FDI in 1993. Closely following China's example is Vietnam, which has fairly successfully adopted a whole series of measures to promote FDI. Since the end of the East European Communist era, most of the East European nations have been actively wooing FDI and they are becoming hot destinations. By the late 1980s, India began to change its traditional attitudes toward FDI and relaxed its rules and regulations. Even hardline North Korea has set aside a small piece of territory along the Tumen River as a special economic zone to attract FDI, though it remains largely closed (Chai 1993: 25–29).

MIXED POLICIES

Most governments have adopted neither a radical policy nor a free-market policy toward FDI, but instead a policy that can be described as a "pragmatic" view, which considers FDI to have both benefits and costs. Therefore, these countries have adopted policies to maximize national benefits and minimize national costs. In order to maximize the benefits, some of them have aggressively courted FDI by offering incentives and subsidies to foreign MNCs in the form of market entry, tax breaks or grants. On the other hand, they have also tended to intervene in FDI in an attempt to limit levels or patterns of investment by MNCs. In a sense, they are caught in a complicated love–hate relationship with foreign direct investment. The restrictions seem to reflect mainly nationalistic positions of countries, even though ideology is an important influential factor in some.

Japan and some EU countries, like France, Italy, Spain, Portugal and Sweden, have traditionally adhered to economic strategies that reserve a wide scope of activity for the government. This statist tradition has consistently been a core aspect of many European and all Latin American economies. Whenever serious problems emerge in private sectors that are considered crucial to national interests, these governments have vigorously intervened to protect local interests. Several major newly industrialized countries (NICs), such as Taiwan and South Korea, are also part of this mixed group, even though they have followed export-expansion strategies. FDI policies of the countries included in this group vary a great deal. To illustrate mixed policies, the national policies of France and Mexico toward FDI are briefly examined.

Owing to a history of governmental guidance for industry arising from Colbertism in the seventeenth century, France has traditionally considered government intervention through ownership or control of business activities as appropriate. With the election of a Socialist government in 1982, the entire banking sector was nationalized to stabilize the exchange rate and control massive capital flight. Although private-sector participation in business has been encouraged, the government has employed "indicative planning" for each sector, backed up by its financial resources. Over half of GDP is generated by the government sector. As governments and the economic situation change, the rules on foreign firms have been tightened or relaxed.

Traditionally, the French government's policies toward MNCs have been heavily influenced by its long-held belief that SOEs are potential "national champions" of its economy. The coal business is dominated by Charbonnages de France, the petroleum sector by Elf-Aquitane, the car industry by Renault, and the steel industry by Sacilor – all SOEs. Major SOE competitors are found in the aircraft industry (Snecma and Aerospatiale), the chemical industry (Rhône-Poulenc and EMC) and the electrical equipment industry (Thomson-Brandt). Foreign firms have been permitted to operate in each of these industries, but their competitiveness has been constrained by the government's support of state enterprises. In the late 1970s, the state was a major shareholder in about 500 companies and held minor interests in another 600 companies (Williams 1987: 80).

Privatization began with the sale of St Gobain in 1986 and Compagnie Financière de Paribas and Compagnie Générale d'Electricité in 1987. This shift toward a greater private-sector role in the economy was not only motivated by the need to rejuvenate French companies' competitiveness. It also partly resulted from the liberalization measures undertaken in the European Union, such as the full elimination of capital controls through the 1988 Directive on the free movement of capital within the European Union. The EU member states most affected by the 1988 Directive were France, which had kept some controls, and Spain and Portugal, which enjoyed an exemption from the full liberalization provisions until the end of 1992 (Thomsen and Woolcock 1993: 69–70).

Nevertheless, domestic ownership is still required in public utilities, national defense industries, stockbrokerage, highway transportation and life insurance. In France, few of the companies listed can in fact be bought. Most of the top forty companies, accounting for 63 per cent of market capitalization, cannot be bought because of family control, cross-shareholdings or state shareholdings (Woolcock 1992: 59–62). Although no local content requirements are imposed on MNCs, contract negotiations to supply the French government often comprise assurances of local production and the use of local suppliers. While no limits on remittance of profits, interest or royalties to a parent exist, price controls are still used in some sectors, notably pharmaceuticals, tobacco and books.

Foreign investors must register proposals with the Ministry of Economy and Finance, though the process is a formality for EU-based companies and does not pose a serious constraint for non-EU firms. Firms from outside the European Union are required to file detailed economic impact estimates and relevant information about proposed projects. The approval process usually takes one month to complete. FDI proposals that are rejected are usually those that plan to enter sectors considered sensitive by the government, or involve acquisitions that do not seem to add any significant financial or technological strength to the French firms. The extent to which the process has hampered FDI has varied with different governments and the economic situation in France.

Among the major Latin American countries, Mexico has until recently had a highly restrictive policy approach toward foreign MNCs, though FDI has managed to develop in the country. With a market size in the region second only to Brazil and a border with the United States, Mexico is an attractive location for FDI. For the past half-century, the Mexican government has played an active part in the economy and laid down extensive regulations for both local and foreign private enterprise. Mexico's 1973 Law to Promote Mexican Investment and Regulate Foreign Investment included rigid regulatory measures to significantly restrict foreign investment (Villarreal 1993: 3).

The Law generally limits foreign investment to 49 per cent ownership, and exclusively reserves certain sectors of the Mexican economy for the Mexican government and domestic investors. But majority foreign ownership (sometimes up to 100 per cent foreign ownership) has often been allowed in certain sectors of the economy, either through decrees governing the automotive, computer and pharmaceutical industries, or through new regulations liberalizing the interpretation of the Law. The Law requires prior approval from Mexico's National Register of Foreign Investment and National Foreign Investment Commission for any new FDI. Permission is given fairly routinely for investment projects that have majority Mexican ownership. Some projects that have been rejected are ones that would compete directly with existing Mexican firms which can serve the market adequately. But if foreign firms seek majority ownership of FDI projects, they have to prove that Mexico can benefit substantially from the investment. The debt crisis beginning in 1982 forced the Mexican government to approve most of the MNC requests to invest, including those with majority foreign ownership, in order to increase the inflow of needed foreign exchange.

The Mexican government is itself heavily involved in the economy, with enterprises in railroads, banking, nuclear energy, petroleum, electricity and telecommunications. The largest company in Mexico is the national oil monopoly, Pemex, which has expanded its activities downstream into petrochemical production. The government development bank, Nacional Financiera, owns significant shares in a wide range of manufacturing enterprises. Foreign firms which have invested in Mexico have found themselves in a relatively unfavorable position when competing with government-supported firms.

Mexico has also imposed various restrictions on FDI projects that seek to market in Mexico. Most products sold in Mexico face some degree of price control, and food and drugs are subject to the strictest control, with price increases allowed only by decree. Many industries are required to use specified percentages of local content in their output. Firms in automobile parts, for example, must use a minimum of 80 per cent. Although trade policy became more open in the 1980s in order for Mexico to join the General Agreement on Tariffs and Trade (GATT), many imports required licenses. Tariffs on fully assembled manufactured goods were usually high, ranging from 20 to 45 per cent *ad valorem*. Tariffs on raw materials and production inputs were comparatively low, or even duty-free (for some materials imported by *maquiladoras*) (Behrman and Grosse 1990: 162). Ironically, this import policy has encouraged FDI to obtain access to the Mexican market.

The restrictions on foreign equity do not apply to foreign-owned, offshore assembly operations (*maquiladora* plants of labor-seeking MNCs), which were first established in 1965 as part of Mexico's Border Industrialization Program. Under the program, foreign firms may set up these assembly plants in Mexico for the purpose of exporting. *Maquiladoras* are permitted to import intermediate materials duty-free with the condition that a significant percentage of the final products is exported. The *maquiladoras* have grown dramatically in the past decade, mainly along the borders with Texas and California. In the early 1990s, *maquiladora* plants in these areas employed over 450,000 Mexicans in a wide range of manufacturing industries for export to the United States (Villarreal 1991).

With the implementation of the North American Free Trade Agreement (NAFTA) beginning in 1994, Mexico's investment environment has been significantly improved. Investors from Canada and the United States enjoy a treatment no less favorable than that which is granted to Mexico's domestic investors. They are no longer subject to any conditions referring to achieved levels of exports of either products or services; maximum national content levels; the granting of preferential treatment to suppliers of domestic products or services; limits on the level of imports to the volume or value of exports, etc. But the new changes do not apply to public-sector purchases and the program of incentives for exports.

Countries in the mixed category present more limitations on FDI entry than do the open ones, but are much less restricted than countries in the restrictive category. The policies of the mixed group seem to become more restrictive as the level of economic development decreases. Governments tend to play a very important role in the economy in terms of policy guide and support of SOEs. Owing to concerns for local capital and entrepreneurs, governments tend to impose restrictions on foreign ownership, though restrictions vary a great deal in different countries and are undergoing liberalization. Price control on various products and some other forms of controls (local content requirements, financial flows, import limitations) are found in the mixed countries to varying degrees. In the 1990s there has been a continuing trend of privatization in many countries in this category. Like the countries in the restrictive category, mixed countries are moving toward more openness to MNCs and decreasing governmental interventions. Regional economic integrations, such as European Union and NAFTA, have served as catalysts for liberalization.

SUMMARY AND CONCLUSION

The three categories of policies of host governments toward FDI are relative divisions, which only describe the main thrust of these policies. The divisions of the three categories do not remain unchanged. In fact, many changes have taken place among the three categories. There is a growing debate in open policy countries, such as the United States, as to whether they should curtail some of their openness. National security and competitiveness have been the major concerns for those who believe that open policies may undermine the national interests of the United States (Tolchin and Tolchin 1988: 15). At the same time, many mixed countries are moving toward open policies. Japan is a notable example. With the introduction of its Foreign Exchange and Trade Control Law in 1980, Japan made a significant move toward becoming a more open economy. Owing to economic reforms and political changes in the 1980s, many countries in the restrictive category have been rapidly moving toward or into the mixed category, including among others China, Vietnam, Russia and most East European nations, and India.

It has become increasingly common for governments to offer incentives to foreign firms to invest in their countries. The most common incentives include low-interest loans, tax concessions and subsidies. These governments are not only motivated by the desire to gain from the resource transfer and job effects of FDI, but also by the desire to lure limited FDI away from other potential host countries. Even local governments of a sovereign nation have actively participated in this competition for FDI. For example, to persuade Toyota to build its US automobile assembly plants in Kentucky instead of other states, the state government offered Toyota an incentive package worth $112 million, including new state spending on infrastructure, low-interest loans and tax breaks (Tolchin and Tolchin 1988). Judging by all the current developments, it is not likely that such a trend of competing for FDI will be reversed in the next decade.

4 National control of technology exports: the case of the United States

INTRODUCTION

As technology is a key for promoting economic development and national competitiveness, less developed countries have been seeking greater access to the technology of industrialized countries. While realizing that export of technology will help expand the export market, many developed countries have shown serious concern regarding national interest implications of unfettered export of advanced technologies. This concern is closely related to both national security and the competitive implications of technology exports. To a certain degree, restrictive measures have hindered international technology transfer. As a major source of technology, the United States has had a long history of controlling the export of military and strategic supplies and technology by enacting various statutes and building up a control system.

There have been, however, two fundamentally different approaches by the US government toward technology export control. One approach has tended to use trade controls to hamper the overall technical advance of US adversaries by restricting virtually any dual-use item that would improve their existing technology. The other approach has advocated the control of only highly strategic items that would directly and significantly promote their military capability (Good 1991: 38). Continual debates around these two approaches have dominated the trends of US technology export control policies. This chapter examines the evolution of US national control policies and systems and discusses their implications for international technology trade.

HISTORY OF US TECHNOLOGY EXPORT CONTROLS

The United States of America pioneered the introduction of control on technology exports with the promulgation of the Export Control Act of 1940. The purposes of this statute were, *inter alia*:

1 To protect the domestic American economy by limiting the export of scarce resources.
2 To direct exports to those countries which would best serve the US national interest.

In 1942, this Act was amended to confer power on the President to prevent exports of "technical data, materials or supplies" (Blakeney 1989: 180–181).

During the first decade after the end of the Second World War, certain trends became clear in the area of export controls. In the Export Control Act of 1949, Congress strengthened and extended wartime powers that authorized the President to regulate exports for reasons of national security, foreign policy or short supply. Congress was prompted partly by postwar economic conditions, but also by concern over the growing military power of the former Soviet Union and the People's Republic of China. To assure the defense of Western Europe, NATO was created in 1949. The successful Soviet explosion of an atomic bomb in 1949 caused more concerns that the Russians had stolen secrets from the United States. Six European nations joined the United States to form the Coordinating Committee for Multilateral Export Control (COCOM), which was designed to facilitate cooperation on strategic embargoes. In 1950, the COCOM member governments drew up the first embargo lists (Melvern *et al.* 1984: 42).

COCOM had no treaty status nor powers whatsoever. Its four basic functions were: arranging agreements on the strategic criteria for export controls; compiling detailed lists of embargoed goods; evaluating individual applications; and coordinating enforcement efforts. All decisions ultimately lay in the hands of member countries acting independently (Hunt 1983: 1285–1297). Since all items within COCOM were kept confidential, including the embargo lists, exporters had to depend on their respective governments to find out what could be sold to embargoed countries. Consequently, different member states developed different interpretations of the rules and enforced varying degrees of compliance (Luks 1987: 98).

Meanwhile, the United States developed its own licensing system as the principal constraint to illegal technology acquisition. The 1949

Act had authorized the administering agencies to attach conditions to export licenses; and the United States used this unilateral authority to exercise jurisdiction over US products and technology which originated in the United States even after they had left American shores. The 1962 amendments made clear that goods of economic importance were to be as much controlled as those of military importance. The amendments also sought to consolidate US controls with multilateral cooperation (Berman and Garson 1967: 791–890). The amendments were not only designed to make export control legislation a better instrument of US foreign policy, but also to influence the foreign policies of other nations.

The use of export controls to serve commercial as well as strategic ends was evident in the Export Administration Act of 1969 (Bingham and Johnson 1979: 894–920). This legislation reflected growing concerns that:

1 US manufacturing enterprises had become relatively inefficient in comparison with those of West Germany and Japan, which were rebuilt after the Second World War.
2 New investments in R&D in the United States had lagged behind those in its principal competitive rivals.
3 The export of high technology, especially in the computer industry, had allowed competitive rivals, through reverse engineering, to enjoy a free ride on the R&D efforts of American enterprises.

The new Export Administration Act removed legal support for economic warfare. The failure of the previous twenty-year embargo to contain Soviet economic development and the increasing reluctance among the US allies to cooperate had a major impact on export control.

In 1979, the United States renewed and amended the Export Administration Act, which established a system for the grant of export licenses by the Department of Commerce. "Validated licenses" were required for the export of:

1 "Strategic" commodities.
2 "Short-supply" commodities.
3 Commodities destined for countries in relation to which there were foreign policy concerns.
4 Unpublished technical data to certain destinations.

"Strategic" commodities were defined in Department of Commerce guidelines as those which were "capable of contributing significantly to the design, manufacture or utilization of military hardware" (Bertsch 1981: 67–82).

The election of Ronald Reagan in 1980 gave a significant boost to export control. The Reagan administration made stemming the flow of advanced technology to the Soviet bloc a high priority by quickly establishing a program to strengthen technology controls. Meanwhile, the US government exerted strong pressure on its allies to comply with US export requirements. Although many of its allies realized the potential threat of the unlimited flow of dual-use technology, they balked at the scope of the American-proposed restriction (Jacobsen 1985: 213–225).

The decision of the US government to insist on rigorous unilateral standards caused an outcry from US businesses, which claimed that they suffered tremendous competitive disadvantages. For instance, prior to trade sanctions against the Soviet Union in 1978, Caterpillar Tractor supplied 85 per cent of the Soviet tractor market. By 1982, Komatsu of Japan controlled 85 per cent of the same market (Lindell 1986: 32). The Export Administration Amendments of 1985 represented a hard-fought compromise between the two schools. These Amendments strengthened the decontrol process for commodities found to be available to the Warsaw Pact.

The growth of the world market and the changing international situation increased the pressure of American businesses for export control reforms. In 1988, Congress required the Department of Commerce to remove the requirement for validated licenses for goods and technology moving to COCOM countries, with exceptions in respect of a few goods such as supercomputers. It also changed the reexport certification requirements to exempt products manufactured abroad using less than 25 per cent of US parts. The 1988 reforms also required the Department of Commerce to decontrol most items determined available from foreign sources in sufficient quantities.

The entire export control system was established as a weapon in the Cold War, to keep free-world technology superior to that of Communist nations. With the end of the Cold War, this system, including COCOM, lost much of the reason for its existence. In early 1994, it was announced that COCOM was to dissolve itself. The US government has also dramatically relaxed control on the export of high technology. In the Spring of 1994, the Clinton Administration announced that it was scrapping virtually all export controls imposed during the Cold War on shipments of telecommunications equipment and computers to China, Russia and most other former Communist countries. The only nations whose civilian companies will remain debarred from access to advanced telecommunications equipment are those on the Department of State list of countries that support or abet terrorism: Cuba, North Korea, Iran, Iraq, Syria and Libya (Friedman 1994: 1). The 1990s will witness a shift in export control from the former Communist nations or reforming Communist nations to the Middle East and parts of Asia in an attempt to control terrorism, the spread of missile systems and non-conventional weapons.

EXPORT CONTROL: NATURE, REGIME AND INSTITUTIONS

As is evident in the above discussion, the US export control system has undergone major changes since the end of the Second World War. Not only have policies changed significantly, but also the responsibilities of the relevant institutions as well as the processes of export control. The following is an overview of the nature, regimes and institutions of the current US export control system (Luo 1994; Long 1989).

Nature of export control

Exports are currently regulated primarily under the Export Administration Act (the EAA) of 1979, as amended in 1985 and 1988. The EAA grants authority to the President, who in turn delegates it to the Department of Commerce. Three primary justifications are offered: to protect national security interests; to promote US foreign policy objectives; and to preserve goods that are in short supply.

National security controls

National security controls restrict the export of goods and technology that "make a significant contribution to the military potential" of other nations, to the detriment of US security. The President compiles a list of countries (controlled countries) that pose a threat to national security by referring to the following criteria: the extent to which its policies are inimical to the US national security interests; the country's Communist status; the present and potential relationship of the country with the United States; and the country's compliance with multilateral nuclear weapons agreements. With the fall of Communist regimes in Eastern Europe and the collapse of the Soviet Union, businesses found it significantly easier to export to those areas.

Countries which are still subject to national security restrictions include Iran and Iraq.

Foreign policy controls

This is the most controversial purpose of export controls. The United States has regularly made a nation's access to goods and technology conditional on its compliance with American foreign policy objectives. The United States has maintained long-term trade embargoes against Cuba, Vietnam and North Korea under the International Emergency Economic Powers Act. The foreign policy provisions listed in the 1979 EAA gave the President the power to impose trade sanctions in response to specific events. For instance, in 1986, President Reagan imposed a total ban on trade with Libya in protest against its support of international terrorism.

Short supply controls

Short supply controls are designed to protect the American economy from the inflation and scarcity that might result from foreign demands for American products in short supply. These occurred during or immediately after major world wars. The Secretary of Commerce is supposed to monitor exports when he suspects they may threaten the domestic economy. When such a threat exists, the President may order quantitative restrictions on exports or a ban on exports of the scarce goods. Under this circumstance, a portion of the export licenses will be distributed to importing countries based on their treatment of US importers during times of short supply. Historically, short supply controls were usually imposed on metallic materials, petroleum products, agricultural commodities, etc.

Major regimes of export control

In addition to the EAA of 1979, there are a number of major export control regimes which have significant impact on the export of technology and are still in effect.

The Trading with the Enemy Act (TWEA) was enacted during the First World War. This legislation prevents US individuals and corporations from exporting goods to enemies of the United States unless they are issued with an export license by the Treasury Department. For forty years, the TWEA gave the President authority to regulate or prohibit exports during any national emergency. Since 1977, it has been used only under wartime conditions, but some pre-1977 TWEA sanctions still remain in force, such as the embargo against Cuba and North Korea.

The International Emergency Economic Power Act (IEEPA) was designed to fill the peacetime gap left when the TWEA reverted back to being solely a wartime measure. It authorizes the President to regulate or prohibit exports whenever he declares a national emergency caused by an extraordinary threat to the national security, foreign policy or economy of the United States from an outside source. During the 1980s, the IEEPA authorized economic sanctions against Iran, Nicaragua and Panama. Currently, trade with both Iraq and Libya is strictly prohibited.

The Arms Export Control Act (AECA) can be traced back to the 1930s, when the United States sought to control its munitions industry. After the passage of the Neutrality Act in 1935, commercial weapons exporters were required to comply with a comprehensive licensing system. The legislation has since been repeatedly amended and is still in force today, known as the AECA. Under the AECA, the Department of State has authority to regulate the commercial export of military-related goods and services with a licensing system. The Act also gives Congress a greater role in weapons export policy, because the executive branch is required to notify Congress at least thirty days before issuing an export license for commercial weapons. The AECA is used to ensure that the export of military-related goods and services do not conflict with US foreign policy or national security interests.

Key institutions of export control

Many international and US agencies are involved in the administration of export controls. Until its expiration in 1994, the Coordinating Committee for Multilateral Export Controls (COCOM) was the most important international agency for export control. Recognizing the ineffectiveness of unilateral controls, it provided members (eventually seventeen) with an opportunity to coordinate their policy making and enforcement efforts. For instance, when a member nation wished to export listed goods or technology to a controlled country, it first sought permission from COCOM. In spite of US pressure, COCOM remained inefficient in exercising technology export control, leaving the real controls in the hands of each member government (Macdonald 1990: 158–160).

In the United States, the Departments of Commerce, Defense,

State, Treasury and Energy are actively involved in the licensing process. The Department of Commerce has authority over the export of materials, equipment and technology that have both civilian and military applications. It delegates most of these functions to the Bureau of Export Administration (BXA). With primary authority over dual-use goods, the BXA is the most powerful institution in the control of exports. The BXA is headed by an Under Secretary of Commerce who is appointed by the President and confirmed by the Senate. Closely related to the BXA are the Offices of Export Licensing (OEL), Technology and Policy Analysis (OTPA) and Export Enforcement (OEE), which provide administrative support for the BXA. The OEL processes export applications according to policies prepared by the OTPA; and the OEE enforces the export controls by issuing subpoenas and making arrests.

The Office of Foreign Availability (OFA) is designed to provide the Secretary of Commerce with information on foreign availability of the goods and technology subject to the export controls. Heading this agency is the Under Secretary of Commerce for Export Administration. The OFA gathers and analyzes a wide range of information in determining foreign availability, with the assistance of US intelligence agencies and private organizations. Foreign availability may be established when: there exists a non-US source for the goods; the foreign goods are comparable in quality to the prohibited US exports; an adequate supply of the foreign goods is available; and the foreign goods are in fact sold to the controlled country. When it is determined that goods or technology are available in sufficient quantity so controls may actually be ineffective, an export license should be approved unless the President deems the issuance of such a license to be harmful to national security.

Licensing decisions concerning defense goods and services lie primarily with the Department of State, which delegates its routine administrative authority to the Office of Defense Trade Controls (ODTC). Manufacturers and exporters of articles appearing on the Munitions Control List are required to register with the ODTC and must not export articles without receiving a license from that agency. In processing its export licenses, the ODTC maintains close coordination with the US Arms Control and Disarmament Agency, which is designed to determine if a particular export might worsen the arms race, help terrorism or undermine world and regional stability. The Defense Technology Security Administration (DTSA) of the Department of Defense is designed to channel the dual-use licensing referrals from the Department of Commerce and munitions referrals from the Department of State to the appropriate agencies within the Department of Defense.

The Treasury Department's Office of Foreign Assets Control (OFAC) was established to administer the Trading with the Enemy Act and the International Emergency Economic Powers Act. It is mainly responsible for trade prohibitions against targeted countries and the freezing of assets of these countries in the United States. In making a licensing decision, the OFAC cooperates closely with the State Department. OFAC is presently administering the sanction regulations against Iraq. Among its past list of country sanctions were China, Vietnam, Cambodia, North Korea and South Africa, to name just a few.

THE EXPORT LICENSING PROCESS

The US government operates a complex licensing system. An exporter's failure to comply with these rules may result in at least long delays and, at the worst, civil and criminal liability.

Type of licenses

All export licenses fall into either of the following two main categories: the general license which does not require governmental approval before exportation, and the validated license which does require written approval from the government before the goods or technology may be shipped out of the country.

Most goods and technology can be exported under a "general license," which allows any person to export low-tech articles to most destinations. Exports under this category can proceed without the need for a licensing application nor any written approval from the government. The exporter should indicate the appropriate general license on its shipping documents. There are more than twenty types of general licenses for commodities. "G-Dest," for example, is used for any commercial goods being exported to any destination under a general license. However, those who wrongly export goods under a general license may be subject to a penalty.

When goods and technology are beyond the scope of a general license, they cannot be exported without first obtaining a "validated license" from the government. Validated licenses must be used within strict time limits and cannot be transferred to similar transactions. There are four basic types of validated licenses:

1 "Individual validated licenses" provide a specific person with a specific power to export a specific item to a specified purchaser on a specific date.
2 "Distribution licenses" permit a specific person to make repeated exports of specific items to a specific destination over a one-year period.
3 "Project licenses" allow a specific person to export all items necessary for the completion of a specified activity for usually a one-year period.
4 "Service supply licenses" permit the export of replacement parts for items that were sold previously for a limited period.

Export classification

In order to find out which type of license is needed, the exporter should properly classify the goods or technology. The relevant criteria applied to this process includes the country of destination, the intended and potential uses for the exports and the identity of the end

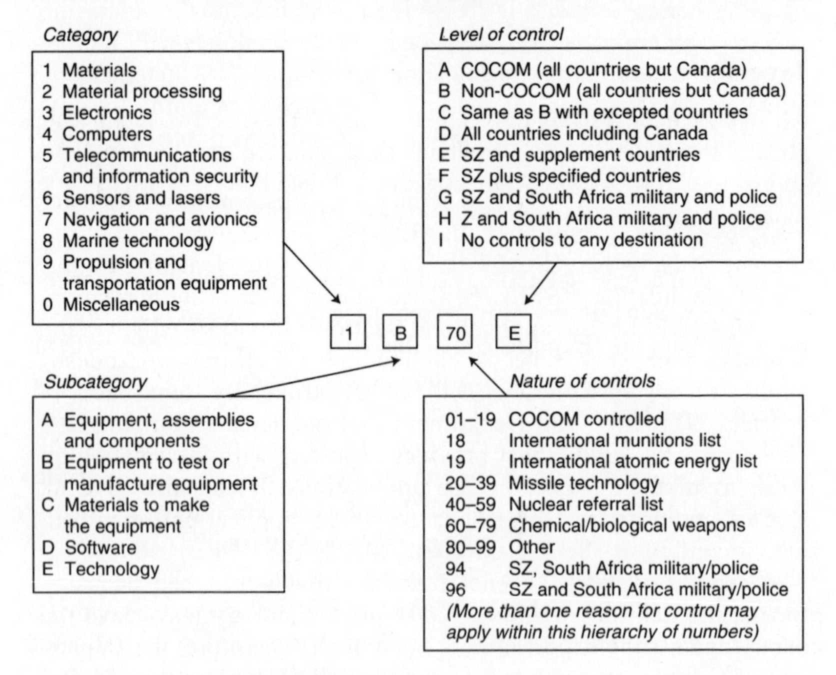

Figure 4.1 A sample of the Commerce Control List
Source: Reprinted from *The OEL Insider*, 3, no. 3, October/November 1991, p. 5

user. To ensure precise classification, one should consult the Commerce Control List to get the appropriate export control classification number (ECCN), which identifies the basic characteristics of the goods or technology, the countries to which it can or cannot be exported and the type of license necessary for each country.

Appropriate documents

The exporter also needs proper documentation to ensure the exports clear US Customs, including the application for an export license, the shipper's export declaration and the import documents. Exporters must fully understand what they are exporting, where the exported items are going, who is going to use them, and how these items are to be used. A validated license generally is issued within ten to twenty days after submission of a properly completed application to the Bureau of Export Administration (BEA), unless the application must be reviewed by another agency. The BEA may also require the exporter to get an "International Import Certificate" from the importing nation, containing the written guarantee of the importing country's government that the goods or technology will not be reexported to a prohibited destination. Sometimes, a "Statement by Ultimate Consignee and Purchaser" is required, containing the foreign buyer's guarantee that it will use the goods as described in the original license application.

Two innovations designed to reduce the paperwork and time for obtaining export licenses are the Export License Application and Information Network (ELAIN) and the System for Tracking Export License Applications (STELA). Once exporters have authorization, they can submit license applications electronically to ELAIN, via the CompuServe network, for most commodities for all free-world destinations. Licensing decisions will be electronically sent back to exporters. STELA is the Department of Commerce's computerized voice answering service that provides exporters with information on the status of their export license applications. It can also give an exporter authority to ship goods for those applications approved without conditions (Smolenski 1992: 20–22).

As exporting under a general license involves a self-licensing process, the risk that sensitive goods and technology may leave the country against the export laws is fairly high. Therefore, the Department of Commerce requires that exports valued at more than $1,500 (or $500 for mailed exports) be accompanied by a "Shipper's Export Declaration" (SED). The Office of Export Enforcement routinely reviews the SEDs for violations. Customs officials may also detain the exported items if there are any discrepancies. All exports under a validated license must be accompanied by an SED.

ENFORCEMENT AND THE TOSHIBA/KONGSBERG CASE

Exporters are expected to develop compliance programs within their organizations. Employees must be educated to recognize "red flags," such as a buyer's hesitation in providing information about end use or the end user; a stated end use incompatible with end user's line of business or destination country's environment; the end user's refusal of the seller's offer of installation services; use of freight forwarders as the ultimate consignees; the buyer's willingness to pay in cash for a large value; the lack of interest among buyer and stated end user in an item's characteristics; and evasive responses to questions regarding any of the above (Macdonald 1990: 138).

The licensing system is also backed up by an enforcement system. Any of the above-mentioned agencies with export control functions may enforce their regulatory policies within US territory. Overseas investigations are also very common. Moreover, US Customs officers are authorized to search, detain and seize items they reasonably suspect are in violation of US export laws. Officers of US Customs and the Office of Export Enforcement have the power to arrest individuals suspected of breaking the law.

Knowing violation of US export laws may result in the imposition of criminal penalties, which may include fines equal to the greater of $50,000 or five times the value of the illegal export, and imprisonment of the incumbent individual for up to five years. The criminal sanctions are much more severe for wilful violations involving exports to controlled countries. Fines up to $250,000 and prison terms up to ten years may be imposed on an individual. A corporation may be required to pay criminal fines equal to the greater of $1 million or five times the value of the exports. Violations may also render the importer liable to civil penalties of up to $10,000 fines (Richards 1994: 322).

In more extreme cases, US enforcement can also be applied to companies from another country in respect of transactions involving non-US technology. A well-known case in this regard is the Toshiba/Kongsberg incident, which took place against the background of growing criticism of Japanese enforcement of COCOM guidelines. Americans were particularly unhappy with the laxity of Japanese companies in exporting high technology to the Soviet Union. The Toshiba incident provided an opportunity for certain high-level quarters within the Reagan Administration to impose sanctions on those Japanese firms which were actively involved in technology transfer to Moscow.

The Toshiba incident can be traced back to 1980 when a Soviet trading company approached Wako Koeki, a small Japanese trading firm, to purchase computerized milling machines and the software to run them. The Soviets wanted these machines in order to fabricate super-quiet submarine propellers that would be far more difficult to detect than regular propellers. Wako Koeki approached Toshiba Machine, a subsidiary of the huge Japanese conglomerate, Toshiba Corporation, which agreed to sell the machinery to the Soviets. Toshiba approached a Norwegian firm, Kongsberg Trading, which undertook to provide the needed computer equipment and software.

In order to obtain the export permit of the Ministry of International Trade and Industry (MITI), Toshiba falsely described the machines and gave a civilian facility in Leningrad as a destination. Four Toshiba milling machines, with nine grinding axes and Kongsberg controls, were shipped from Japan to the Soviet Union between 1981 and 1983, and a further four (with five axes) were shipped in 1984. The export of such milling equipment contravened COCOM guidelines. The machinery enabled the Soviets to build submarines that could avoid detection by US defense systems. Thus both Toshiba and Kongsberg broke their own national laws and were to be punished by their own governments. Investigations were under way in both countries.

The US Congress reacted with outrage to the sale of such sensitive technology to the Soviets. As the exports did not originate in the United States or use US technology, the two incumbent companies were not subject to US export control laws (Wrubel 1989: 241–261). Partly because of this, Congress amended the Export Administration Act in 1988 to allow the President to impose sanctions against foreign firms violating national export controls imposed pursuant to COCOM regulations when those violations compromised US national security interests. The Congress also imposed a three-year ban on imports and on US government procurement against Toshiba Machine and Kongsberg Trading, and a three-year ban on US government procurement from Toshiba Corporation and Kongsberg Vaaperfabrikk, the parent companies.

SUMMARY AND CONCLUSION

This chapter has reviewed and analyzed the history of US export controls, the nature of US export control, and existing regimes and institutions on export control as well as the export licensing process. The issue of enforcement has been illustrated by the Toshiba/Kongsberg case. Not only is the system complicated and difficult to work with, but it also has been a source of friction between the United States and its Western allies. There has been growing pressure for change from both US allies and American businesses, which have allegedly been placed in a relatively disadvantaged position to compete in the global market. A review of its history since the end of the Second World War shows that such control has been anything but efficient. It has also had an unfavorable impact on technology development in the West. As Stuart Macdonald summarized the situation:

> The supreme irony of export controls, though, is that they are justified in terms of insuring the West's military strength through technological supremacy: it is difficult to imagine any artifice more likely to undermine the West's capacity for technological innovation than the current systems of national security export controls.
>
> (Macdonald 1990: 201)

With the end of the Cold War, US technology export control is undergoing fundamental changes. Economic interests are gaining significant weight while strategic and security interests are losing their traditional dominance. The Clinton Administration has found it necessary to keep a good balance between preventing the proliferation of dangerous weapons and sensitive technologies on the one hand, and ensuring that American companies remain the most competitive in the world on the other (Friedman 1994: 1). As global technology advances, export controls must be updated in order to remain focused on those items that directly cause proliferation concern. The fact that the licensing requirements for many high-tech products, such as most civilian telecommunications equipment and computers, have been removed, is evidence of the dynamism of such a trend. Continued improvements in license processing will further reduce the impact of export controls. Under the new international environment in which excellence in economic competition is becoming a national priority of most governments, national technology export control is becoming an increasingly unpopular liability.

Part II

International protection of intellectual property

5 The international dimension of intellectual property

INTRODUCTION

Intellectual property law has always had an important international dimension. The need for international treaties protecting intellectual property resulted in the Union of Paris for the Protection of Industrial Property (1883) and the Berne Convention for the Protection of Literary and Artistic Works (1886). The international dimension of intellectual property is now becoming even more important for three compelling reasons (Leaffer 1990: 1–2). First, the composition of world trade is changing, with trade in intellectual property becoming a greater part of commerce between nations. The value of informational products has been significantly increased by the new technologies of the semiconductor chip, computer software and biotechnology. Second, global commercial relations have become ever more interdependent, creating a greater need for international cooperation. More and more countries have recognized the increasing need for international agreements involving intellectual property. Third, new reprographic and information storage technologies have enabled unauthorized copying to occur more efficiently than ever and significantly devalued creative work. Thus, there is a growing need to revise traditional international treaties as administered by the World Intellectual Property Organization.

Traditionally, people have thought of property as either movable property (e.g. a radio or a bicycle) or immovable property (e.g. a house or land). One characteristic of these forms of property is their tangible shape. In contrast, intellectual property law confers property rights on intangibles. Intellectual property can be characterized as "pieces of information which can be incorporated in tangible objects at the same time in an unlimited number of copies at different locations anywhere in the world. The property is not in those copies but in the information reflected in those copies" (WIPO 1988: 3). Intellectual property is generally divided into two major branches: industrial property and copyright law. Industrial property includes patents, trademarks and industrial designs. Copyright law and neighboring rights cover artistic, musical and literary works. No international treaty completely defines these types of intellectual property, and the laws of countries differ from each other in significant ways. This chapter reviews some fundamental concepts of intellectual property rights.

PATENTS

A patent can be defined as a statutory privilege granted by a government to an inventor and to other persons deriving their rights from the inventor, for a fixed period of years, to exclude other persons from utilizing a patented technology (method and process) (UN General Secretary 1972: paragraph 1).

The idea of a patent is to protect an inventor so that he or she can profit properly from the invention, thus providing an appropriate incentive for innovation. A legal monopoly on the patented technology is granted. A patent covers an invention which is novel, considerably different from presently known technology, is non-obvious, and has commercial value. The legal validity of a patent may depend on the right of prior discovery in a group of countries or registration in many others, or both in some countries. The most significant device for protecting internationally transferred technology is probably the patent. One needs first of all to obtain a valid patent in the country from which the technology originates and then to assure enforcement of that patent in the recipient country. The term of patent protection varies from five to twenty years.

As a common practice in most countries, the applicant for the protection of an invention should file his or her application with the relevant industrial property office. The application should contain, among many other items, a description of the invention, with any drawings referred to in the description, and the claims made for the invention. The description should disclose the invention in a manner sufficiently clear and complete for it to be carried out by a person having ordinary skill in the art. The disclosure of the invention should present the invention in the context of the state of the art (WIPO 1979: section 112(1)).

Most countries grant patents only after the invention or product for which protection is sought has undergone three tests. Under the test of novelty, the invention must not have been published in any place, domestically or abroad. The United States accepts a patent application at any time up to one year after publication, but many countries do not. Most European nations, for example, require that a patent application be filed before the invention has been publicly disclosed anywhere in the world, by any means, written or otherwise. This is called the "absolute novelty" requirement. Although the information may have been disclosed only in the United States, the inventor loses European patent rights if he has not applied for a patent first in Europe. It may only require communication to a potential customer for the disclosure to be deemed "public." Almost all major countries in the world follow to varying degrees the absolute novelty rule. Only Canada and Russia have similar novelty requirements to that of the United States in that an inventor has a one-year grace period following his own public disclosure or disclosure by someone who learned of the invention from him and made it public without authorization (Curesky 1989: 289–308; Behringer 1994: 58–59).

The test of novelty also means that no one else has come up with the same or similar idea (Robinson 1988: 133). This requires that a search should be made by the relevant patent office; in some countries this involves a time-consuming and thorough process, taking up to several years, while in others the process is much simpler and shorter. The products of the former are referred to as "strong patents," which are difficult to challenge effectively; and the latter as "weak patents," the challenge of which are relatively easier. In fact, one may not know if the patent protection granted will hold or not until after a certain period of time has passed without effective challenge to the patent. Patents granted by the United States and Germany are generally deemed as strong while those awarded by countries such as France, Italy, Luxembourg and Belgium, as relatively weak.

The second test usually applied to a patent application is that of non-obviousness. The tricky issue is to determine what people are covered under non-obviousness. Should the invention be non-obvious to ordinary people in the street? Or should it be non-obvious to a reasonably well-educated person knowing the relevant technology? For the purposes of most patent laws, an invention is considered to involve an "inventive step" if, having regard to the prior art, "it would not have been obvious to a person having an ordinary skill in the art" (WIPO 1979: section 114(2)(a)). In other words the invention must represent a creative leap forward on existing knowledge.

The third test is associated with commercial value: the invention qualified for a patent should have an industrial applicability. This test excludes from patent protection purely theoretical inventions that cannot be put into industrial production, thus having no commercial value. The notion of an industrial applicability implies a commercial scale of application, though industry is hereby used in its broadest sense to embrace also extractive industries, agriculture, fishing and forestry (WIPO 1979: section 116).

There are various restrictions on the grant of patent and the transnational movement of patents (Baxter and Sinnott 1983). The United States, for example, prohibits the foreign filing for an invention made in the United States, without first securing a foreign filing license from the US Patent and Trademark Office, unless six months have elapsed since filing a US application. This rule is designed to protect against transfers of technology possibly damaging to national security. Violations of this requirement can result in either loss of the US patent right or a criminal penalty or even both. Algerian law requires that a patent of importation can be issued only where the importance of the invention, the industrial or agricultural advantages, and the public usefulness of its exploitation warrant such protection.

Some countries refuse to award patents for products and/or processes relating to certain types of technology. For reasons of national interest, some countries withhold patent protection from inventions relating to agriculture, food, medical and pharmaceutical products and nuclear technology. In the United States, for example, certain inventions in the atomic energy area, antipollution devices, methods of doing business, and medicines that are more in the nature of mixtures, may not be awarded with patent. Under Mexican law, patents will not be granted to inventions pertaining to food processing, pharmaceutical chemicals, agricultural fertilizers, pesticides, herbicides, pollution control or nuclear energy. Sometimes, the product itself may receive patent, but the processes are excluded or vice versa.

Many countries require the licensing of a patent if they are convinced that local manufacture is not sufficient to meet local demand. While this system of "compulsory licensing" is fairly common in British Commonwealth countries, many developing countries have also maintained such a practice. It is not uncommon that the foreign patent holder in one of these countries has two or three years of grace, in addition to two more either to work the patent locally by using it in the holder's own facilities or to license it to another party. In some Latin American countries, foreign patent holders are required to take constructive steps to get second parties interested in taking a license under a patent. Otherwise, they can lose patent protection in these countries. Some countries grant "confirmation patents," to the owners of foreign patents, if they apply within a certain period. Other countries may issue "patents of introduction," which are short-term patents for something already patented elsewhere. Like "compulsory licensing," these patents are introduced to encourage the use of the patent locally (Robinson 1988: 134).

Despite various international agreements, patent systems can be discriminatory against foreign applicants, as is evident in a study by Kotabe (1992: 147–168), who found that Japanese patent practices discriminate against foreign applicants with longer pendency periods than for domestic applicants, while the US, British and German systems discriminate against foreign applicants with lower patent grant ratios than for domestic applicants. He also indicates that as long as foreign applicants know the system well, they can take full advantage of various loopholes inherent in the Japanese system. For instance, a growing number of translation businesses that cater to American firms have been getting access to Japanese patent applications published within eighteen months of filing. Such access keeps American firms abreast of innovations originating in Japan, subsequently allowing them to practice "patent flooding" in Japan by filing many improvements around the original Japanese inventions. This has been practiced by Japanese firms against their Japanese competitors for years.

INDUSTRIAL DESIGN

An industrial design refers to the ornamental or aesthetic aspect of a useful item. The ornamental aspect may be embodied in the shape, pattern or color of an item and it must be able to be reproduced by industrial means. The protection for industrial designs can be obtained through registration. The beneficiary of registration is granted with either a design patent or a certificate of registration (WIPO 1970: section 2).

As with patent approvals, most countries which protect industrial designs require novelty. The criterion of novelty varies a great deal between universal novelty or merely national novelty. A complicated problem in design protection is the extent to which a design must differ from an earlier design to be considered novel. Minor variations are usually considered insufficient to qualify an existing design as novel. A widely accepted test rule is whether the design claimed is subjectively new in the sense that it is not deemed as an imitation of designs already known to the creator (WIPO 1970: section 4(4)). The key qualifying feature of industrial applicability is that the design is repeatable in commercial quantities. Therefore, items of artistic craftsmanship do not fall within the scope of design protection and are more commonly covered by copyright law.

In some countries, protection may be granted for certain objects of utility which embody a more limited technological advance than that required for patentability. The protection normally covers a more limited scope and period than that awarded by a patent. However, the patent office requirements are not as demanding as those for patents. This type of protection is designed to bridge the gap between patent and design protection. Nevertheless, it is also an important incentive to inventive activity in developing countries where initial inventive activity will invariably contribute minor additions to existing technology. Industrial property statistics show that in those countries which provide for the protection of utility models, the great majority of applicants are nationals or residents (UNCTAD 1975: 40).

BIOTECHNOLOGICAL RIGHTS

Biotechnological invention, especially through the practice of genetic engineering, has been a highly controversial issue. The issue of plant breeders' rights can be traced back to the 1920s, when a number of European countries recognized various kinds of plant breeders' rights. Since the 1930s plant varieties have been included in patent protection in the United States and Germany and later on in Austria, Belgium, the Federal Republic of Germany, France, Italy, Hungary, Japan and Sweden. In 1961 an International Convention for the Protection of New Varieties of Plants was concluded in Paris.

In comparison, the question of the patentability of "animate" substances had been handled in a separate legal channel from the issue of the patentability of plant strains. Originally it was widely accepted that discoveries pertaining to living organisms and material were not "inventions" for the purposes of most patent statutes, though there were some notable exceptions, including among others micro-organisms used in fermentation and antibiotics. The Supreme Court of the Federal Republic of Germany created the precedent of protection when it ruled in 1969 that animal breeding techniques were patentable, provided that the technique was repeatable. Still, the US courts repeatedly rejected claims for the patentability of animate matter until the 1980 decision of the Supreme Court in *Diamond v. Chakrabaty* (447 *US* 303 (1980)). In that case, the Court decided that a genetically-engineered bacterium capable of breaking down the components of crude oil could be patented. The case established a principle that the critical standard of patentability was not whether an invention pertained to living or inanimate objects, but whether it involved a human-made invention (Blakeney 1989: 12).

The reason for denying protection to biotechnological inventions in many developing countries has been quite different, as many of them believe that food and health are so important to their economic development and the basic livelihood of people that they should not be monopolized by private firms. Their weak protection of biotechnological rights is often excused as a way to avoid restrictions in "the supply of essential products" (Twinomukunzi 1982: 31–68). This category of invention is clearly excluded from protection by a number of patent statutes. Since the 1970s, there has been a growing demand among developing countries for access to biotechnology, though the reactions from developed countries were skeptical and apathetic. By the 1980s, when environmental problems began to be a primary concern of the developed countries, the developing countries pressed for greater access to world genetic resources by linking this issue to the protection of the environment.

The Earth Summit in Rio de Janeiro in June 1992 was a good example in showing the major difference between the developed and developing nations on this issue. President Bush refused to sign the biodiversity treaty on the ground that it would subject American biotechnology companies to compulsory licensing. The aim of the treaty was to ensure that countries curb the destruction of species, habitats and ecosystems. The treaty requires the United States and other developed countries to endorse the principle that when a gene is used from an organism found in another country, that country is entitled to compensation. It is designed to give developing countries sufficient incentives to protect their biological diversity, including plants and animals in rain forests. The treaty received overwhelming support from developing countries (*CMR* 1992: 3).

TRADEMARK

A trademark represents a certification of origin, thereby telling the customer something about product quality. Trademark as a property originally came from the craftsman's distinctive mark which identified the item with a particular craftsman, whose reputation could add value to the product. As with patent, the validity of a trademark (or trade name) rests with either prior use in trade and/or registration. To be registered as a trademark a sign must be visible rather than audible or olfactory. Registrable signs include names, existing or invented words, letters, numbers, pictures and symbols (Blakeney 1989: 11). In some countries, the shapes of goods or their containers may be registered as marks.

A trademark may be a valuable asset if it is well established in the trade. Its use may facilitate market entry of a product or service by reason of greater consumer acceptance. Trademark covers four different categories (Robinson 1988: 135):

1 The "true trademark," which relates a product to a specific firm.
2 A service mark, which identifies a service (a "soft" product, like the name "Arthur D. Little").
3 Certification marks, which represent a certain quality (for example, Underwriters' Laboratory stamp of approval).
4 Collective marks, which refer to those used by a group or organization to distinguish the characteristic features of products used by that group or organization (such as membership in an association).

Unlike a patent, protection of a trademark may be indefinite as long as basic conditions are met.

Excepted from registration in most countries are marks which are not distinctive or are deceptively similar to existing marks and those which violate public order or morality. Generic terms are normally not qualified as trademarks. A mark can phase out if the word becomes part of the common vocabulary (such as aspirin, cellophane, cola, escalator, kerosene, nylon, thermos and zipper). The Federal Trade Commission of the United States is charged with the responsibility to determine if a trademark has become a generic term for a certain category of goods or services to be usable by anyone. Marks should not be created to be confusingly similar with those already existing. A sportsbag made in China with a "NIKF" mark is too similar to the NIKE brand for a Chinese consumer to make the distinction (*Financial Times*, January 10, 1995: 6). A different language or culture may have unfavorable or even rude meanings for words and symbols with favorable meanings in the home country. White, for example, may imply purity in the United States, but is the color of mourning in most of the East Asian countries.

Common law countries, including the United States and members of the British Commonwealth, usually require continuous prior use in trade by the applicant as a pre-condition for issuing a valid trademark registration. Continuity of prior use, even if not registered, may provide adequate ground for protection. These marks are known as the so-called "common law trademarks." Owing to the constitutional restraint on federal powers in the field of commerce, the US government can issue a trademark only if it has been used in interstate or foreign trade. However, trademarks registered elsewhere, whether already used or not in US trade, are recognized by the US government. Civil law countries (most other countries in the world) provide protection mainly on the basis of the act of registration. Consequently, a firm may lose its trademark if it fails to register it and to maintain its validity by paying an annual fee. Demonstrable prior use may not be sufficient for restoring ownership.

For those companies which are not quite ready to do business abroad, it is prudent to decide early where trademark protection will be needed and to protect rights by filing in those countries. The question of where to file is essentially a business decision, which entails balancing the expense of registration against its potential benefit. At least, one should file in countries in which one will do business. Companies should also file in countries which are known sources of counterfeit goods, though some countries may require local use to maintain a registration. While trademark laws do not normally impose a deadline for registering a mark, a company which owns a reputable mark should register promptly in the country of potential interest or stake so that other companies will not be able to register the mark in that country (PTOUS 1991: 5).

Closely related to trademarks are trade names and appellations of origin. Trade names are the names under which business enterprises trade. In many countries, they can be registered as trademarks. But in some countries, trade names are set aside exclusively for a registering enterprise. An appellation of origin is a geographic location which serves to designate a product. There is a consistent and unique connection between the location and characteristic features of the product. This is particularly true for many agricultural products. Cashmere wool and Bordeaux wines, for example, are products which owe their distinguishing features to the areas in which they are cultivated.

TRADE SECRETS AND KNOW-HOW

A trade secret is know-how that is kept secret within a business, offers a competitive advantage, and is not generally known to the industry. Trade secrets cover manufacturing processes; methods and techniques; plans, designs and patterns; formulas; business information; and product.

Trade secrets do not enjoy statutory protection like patents and trademarks, but the laws of many countries provide some sort of protection to ownership rights in trade secrets, conditional upon the maintenance of secrecy. It is not always possible to keep the secret, because a competitor can use a variety of channels to get the secrets, including hiring personnel and reverse engineering. Once a trade secret is publicly known, anyone can use it freely. Trade secrets do not give the owner any right to prevent others from independently creating and using the same know-how.

Companies may not be interested in applying for patent protection for a variety of reasons, including the time and cost involved in patenting, the uncertainty of patent validity and the difficulty in monitoring patent infringement and enforcing patent rights against alleged infringers. There have already been some statistical reports that well over half of all patents issued in the United States are ultimately rendered invalid. Trade secret protection may exist forever if nobody is ever able to independently find out the secret. Coca-Cola is a typical example of a company which has resorted to the trade secret protection of its "ingredient X." The formula for Coca-Cola has remained a trade secret for more than 100 years. If the formula had been patented, it would have been made available to the public within seventeen years (Robinson 1988: 138).

Under the laws of some countries, it may not be too difficult for a company to prevent employees from committing trade secrets abuses or to prevent a competitor from utilizing a trade secret acquired illegally. In France, for instance, clauses prohibiting employees from communicating trade secrets without authorization are often contained in collective labor agreements in a specific industrial sector. To prevent a third party from wrongfully acquiring and using a person's trade secrets, courts in the common law system have specified rights in tort and equity. In some legal systems, the unauthorized incorporation of confidential information in innovations of the wrongful user is prevented by the "springboard doctrine" (Blakeney 1989: 13–14). According to the doctrine, a person who has obtained information in confidence is not permitted to use it as a "springboard" for activities harmful to the person who made the confidential communication. A person who wishes to make use of confidential information under these circumstances is required by law to pay damages for the privilege.

Corporations operating internationally may also suffer from unusual access to sensitive corporate information by other entities. This problem is partly related to the personal privacy law that most countries have. These privacy protection laws allow citizens (employees, customers, etc.) on whom a corporation may be keeping information to access the information kept by the corporation. But the personal privacy law itself does not necessarily allow access to corporate proprietary information. Only when countries define persons to include "legal persons" may there be a serious problem. Countries which do have such an interpretation include Austria, Denmark, Germany, Luxembourg, Norway and New Zealand. Therefore, a country's privacy statutes should be carefully studied and appropriate precautions should be taken in regard to files kept at the overseas operations (Ewer 1993: 46–48).

In short, the general lack of international treaties, in addition to the costs and uncertain outcome of legal action, make it crucial that a licensor take the best possible contractual safeguard measures to prevent the disclosure or unwarranted use of trade secrets by a licensee before, during and after a licensing agreement. Most developed countries have mechanisms to enforce a written agreement not to divulge a trade secret, but many developing countries neither recognize nor enforce rights based on trade secrets. Although many of them do have some rules against unauthorized communication and use of trade secrets, enforcement has been very weak, or in some cases almost non-existent. This has considerably increased difficulties and risks of international technology transfer. Therefore, it is highly recommended that before divulging any information which may be useful for a potential competitor, one should investigate the protection available in the recipient's country.

COPYRIGHTS

Copyright protection covers the works of authors and artists, by giving the copyright owner the right to control the reproduction and performance of the work. While the scope and length of copyright protection varies a great deal among countries, copyright generally provides protection for written works, films, music and performances. Most copyright laws protect works for a significant period of time, such as the life of the author plus fifty years. Because copyright laws vary considerably throughout the world, each nation's laws must be investigated individually to understand the level of legal protection available for copyrighted works.

In many countries (including the United States), a copyright is automatically endowed to the creator on the creation of the work. Although owners may feel more secure if they register their copyrights with the appropriate governmental office, they are not required to do so. Copyright owners usually serve notice of their copyright by indicating its existence on copies of their works. In the United States, a copyright infringer does not have adequate excuse to reduce his or her liability by claiming innocent infringement if appropriate notice has been provided.

Copyright laws have followed two different approaches in their development (Hotchkiss 1994: 305–306). The common law legal system tends to consider copyrights as a protection solely of economic interests. But for civil law countries, copyright is supposed to protect both the economic and moral rights of authors. Under civil law systems, owners of copyrights have to recognize the rights of the original authors. For instance, someone who has purchased a painting by Van Gogh and thus owns it may not have the right to cut it into small pieces, selling each as an "original Van Gogh." Such an act is viewed as a violation of the artist's right to maintain the integrity of his work.

The significance that civil law systems have attached to the moral rights of authors should not be ignored. Moral rights refer to rights that stay with the author even though the work may be sold to someone else. Generally speaking, moral rights comprise two parts. The first part connotes the right of paternity: i.e. the author's right to be identified with the work. The second part refers to the right of integrity: i.e. the author's right to retain some control over what subsequent owners do with the original work. The right of integrity influences such upgrading efforts as "colorizing" old black-and-white movies, as well as other republication or distortions of the work that would tamper with the author's original creative thought.

Some recent developments in Europe have caused concern to US copyright owners (O'Connor 1993: 36–38). Several EU member states are creating new rights and classes of right holders that are not covered by international copyright rules. US copyright owners charge that these new classes of right holders are making unfair profits from the dissemination of copyrighted works. A typical example is a videogram producer right in France. A videogram producer is the individual who first fixes the cinematographic work in videocassette form. Unlike the United States, France does not consider the rights of the videogram producer as governed by copyright protection. Under the French laws, the rights of the videogram producer should be covered by so-called neighboring right rules. Since neighboring rights are governed by the reciprocity principle instead of national treatment contained in copyright, the United States does not recognize such rights. For France, therefore, there is no obligation to the United States in this area.

For the past few decades, the technological advances brought about by computers have created special problems for intellectual property protection all over the world. Neither computer software nor hardware fits well with the framework of copyright or patent law. Although computer software is generally protected under copyright law, the Berne Convention was signed decades before the widespread applications of computers, and therefore has no provision on computer software being subject to copyright. Therefore, a software owner will have to examine each country's copyright law to see if it covers software. Many countries have already viewed software to be copyrightable, but others have not yet taken the issue seriously, or refuse to grant proper protection.

Computer hardware, particularly chips, has been even more controversial. In 1984, the United States established a new class of intellectual property for chips. The Semiconductor Chip Protection Act of 1984 created a special system of legal protection for original mask work used in the production of semiconductor chips. It confers the exclusive right to reproduce and distribute mask works for a term of ten years from registration. The Act was introduced to provide other legal systems with incentives to create similar mask work protection. For foreign chip designers who seek US protection, the Act requires reciprocity. If the foreign nation protects mask works and extends that protection to US chip designers, the foreign designer may enjoy protection for mask works in the United States (PTOUS 1991: 5). The Act has proved to be reasonably successful in inducing other legal systems to enact legislation protecting mask works. The European Union, Japan and some other nations have passed such laws. Countries have become increasingly interested in a multilateral agreement that would protect mask works internationally.

SUMMARY AND CONCLUSION

This chapter has introduced basic concepts of intellectual property rights, ranging from patent, industrial design, biotechnology rights, trademark and trade secret to copyright. Ideas have brought the world waves of technological revolutions that have significantly changed the process of production and the ways of life and work of human beings. The personal computer revolution, for example, has revolutionized almost every aspect of work and life. Various new products, such as new drugs and high-tech gadgets, have been introduced to the global market at an unprecedented speed. The legal protection of intellectual property rights is becoming increasingly important to the continued dynamic development of these industries.

Over the past few decades, significant progress has been made in the area of intellectual property protection. The necessity to protect these intellectual property rights has been generally accepted in the world, but major differences in approach have remained not only between developed and developing countries but also between developed countries. These differences, if not properly understood, may cost businesses operating internationally hundreds of millions of dollars. To enhance the protection of intellectual property, there is still a great need for harmonization of these various laws and rules.

6 International protection of intellectual property

INTRODUCTION

Intellectual property rights conferred by the various intellectual property laws discussed in Chapter 5 generally have no extraterritorial effect. Before the establishment of the Paris Convention for the Protection of Industrial Property in 1883, foreign protection of intellectual property rights was dependent on reciprocal recognition granted by the laws of various countries. Since the signing of the Paris Convention, however, a number of international conventions, treaties and agreements have been promulgated to promote and harmonize the principle of reciprocity in the field of intellectual property. The most important of these international agreements in addition to the Paris Convention are the Patent Cooperation Treaty, the Madrid Agreement, the Berne Convention and the Universal Copyright Convention. Although these agreements do not preempt national law, they have provided general guidelines that have facilitated the international transfer of technology.

The establishment of these various international conventions, treaties and agreements have benefited both developed and developing countries. These agreements are important for developed countries, as a rapidly growing percentage of their exports depends on proper protection of intellectual property. Participation by developing countries in these agreements also facilitates the dissemination of information conducive to the development of a global legal infrastructure for the protection of intellectual property. The principal advantage for developing countries rests with the fact that these agreements have provided a favorable international framework for the transfer of developed intellectual property expertise to developing countries. The applications of the various intellectual property classifications are beneficial to these infant intellectual property protection systems.

Single registration systems embodied in some of these agreements provide the developing countries with badly needed search resources and expertise of the central registration agencies. The World Intellectual Property Organization (WIPO) has facilitated the implementation of these various agreements.

PARIS CONVENTION FOR THE PROTECTION OF INDUSTRIAL PROPERTY

The Paris Convention, signed in 1883 and revised at Brussels in 1900, Washington in 1911, The Hague in 1925, London in 1934, Lisbon in 1958 and Stockholm in 1967, established an international union for the protection of industrial property (the Paris Union) (Leaffer 1990: 17–49). It has now been signed by over 100 countries of the world. Accession to the Paris Convention is effected by the deposit of an instrument of accession with the Director-General of WIPO, who is also the Chairman of the International Bureau which administers the day-to-day operations of the Paris Convention.

The principal rules of the Convention may be divided into four categories: substantive rules of law guaranteeing the right of national treatment; rules of law which guarantee the right of priority; clauses pertaining to specific topics of industrial property law; and principles for the establishment of an administrative framework for the implementation of the Convention. The first two constitute the most important principles of the Convention (Leaffer 1990: 17–49).

The national treatment principle, which is expounded in Article 2, provides that members of the Paris Union are required to grant the same industrial property protection to nationals of other member nations as it grants to its own nationals. "Consequently," Article 2 provides, "they shall have the same protection as the latter, and the same legal remedy against any infringement of their rights, provided that the conditions and formalities imposed upon nationals are complied with." No requirement regarding domicile or establishment in the country where protection is claimed may be imposed upon nationals of member nations of the Union as a condition of entitlement to any industrial property rights.

Nevertheless, the prohibition of discrimination contained in Article 2, either against foreign nationals or in favor of indigenous companies, has been criticized by many as being unfair to developing countries. Various proposals have been forwarded to restructure this legal framework, including the waiver of the national treatment principle for developing countries. It is argued that:

> formal equality as provided for by Article 2 would operate to the mutual advantages of the Convention countries if they were either at or almost at the same level of technological and economic development. However with the immense diversity in technological capabilities between the developed and less developed member countries, the principle simply confers on the more developed members the unlimited rights to the detriment of the other.
>
> (Yankey 1987: 63)

The right of priority, which is spelled out by Article 4 of the Paris Convention, expounds that on the basis of an application for an industrial property right filed by an applicant in a member country, the same applicant may, within a specified period of time, apply for protection in all other member countries. These later applications will be treated as if they had been filed on the same day as the first application. The priority period for patents is twelve months. The priority principle in the case of patents connotes that an applicant for registration in a member country should be protected from the loss of novelty by any exploitation of the patent during the priority period in any other member country.

The priority principle grants innovators the necessary time to apply for protection in other countries. But many think that the period covered is exceptionally long and regard the principle as a disincentive to the exploitation of new inventions in developing countries. With the dramatic improvements in communications since the Paris Convention, there is already a strong argument that its priority periods have become excessive (Alikhan 1993: 219). The priority principle may also cause some confusion for applicants from common law countries. Under the Paris Convention, the right to such priority has to be specifically claimed with a certified copy of the first application. But if the US application, for example, is not filed until after a public disclosure, the Paris Convention can be of no use. The holder of a valid application for a US patent may lose foreign patent rights if he did not apply early enough in the USA. Although filing during the grace period obtains US rights, it is necessary to file before the grace period to preserve foreign rights (Behringer 1994: 59).

There are specific provisions in the Convention that deal with patents and trademarks. As far as the transfer of technology is concerned, the most important of these are probably those associated with the compulsory licensing of industrial property. For instance, where a patent is considered to have been insufficiently worked for a specified period of time within a country, Article 5A of the Convention allows either the compulsory acquisition of a patent or its compulsory licensing to another company. This provision is created to impede the abusive practices of the anti-competitive hoarding of patents. The compulsory licensing of patents may be applied to other abuses, including excessive prices and restrictive terms in patent licenses. Similar provisions are contained in Article 5C of the Convention with regard to the non-use of trademarks.

The Paris Convention was last revised at Stockholm in 1967. Although there is a growing need for more improvements, the prospect of replacing the Act of Stockholm seems to be dismal for a number of reasons (Sinnott 1993: 31–36). Some of the intellectual property issues that remained unresolved under the Convention revision negotiations were routed into the General Agreement on Tariffs and Trade (GATT), Uruguay Round negotiations. Moreover, substantial progress has already been made in the direction of establishing a modern, international standard for intellectual property law, which is embodied in the proposed Patent Law Treaty, the proposed Trademark Law Treaty and the proposed Protocol to the Berne Convention. Finally, the Groups of Developing Countries' interest in greater flexibility in compulsory licensing rules under the Convention also poses a major hindrance to the negotiations.

PATENT CONVENTIONS AND TREATIES

Apart from the Paris Convention, there are a number of important international agreements specifically covering patents. The most significant of these include the Patent Cooperation Treaty and the Strasbourg Agreement.

Patent Cooperation Treaty

The Patent Cooperation Treaty (PCT) was adopted in 1970 at a diplomatic conference in Washington for the purpose of simplifying the process of securing international protection for patents. It was amended in October 1979 and modified in February 1984 (Leaffer 1990: 76–126). More than forty countries (including the United States) are parties to the Patent Cooperation Treaty.

Before the introduction of the PCT, a patent applicant had to file an application for registration in all the countries in which the applicant intended to apply the invention. This required the applicant to comply with the disparate filing procedures of each country and to meet with the search requirements of those countries. The PCT provides an applicant from a member nation with the mechanism to file a single application for patent registration that designates the member nations in which the applicant seeks protection. After this filing an international search is undertaken by a designated searching authority to determine the patentability of the applicant's invention. The application and search report are then dispatched by WIPO to all relevant patent offices.

In addition, the PCT provides a facility for a preliminary international search of an invention's novelty, obviousness and industrial applicability. This search evaluates whether it is worth the expense of proceeding with an application. Such a facility is very beneficial to developing nations, which generally lack resources for the costly and complicated searching required for patent registration. An international application can also be filed in the applicant's own language, with the applicant's home patent office acting as a receiving office under the PCT, and fees paid in local currency. A translation is not necessary until entry into the national phase of processing.

The fact that the PCT procedure is particularly attractive to applicants is apparent from the annual growth rate in filing applications, which in the past few years has consistently surpassed 25 per cent. In 1990, more than 19,000 international applications were filed, having the effect of more than 400,000 national applications. Since the PCT came into force in 1978, more than 100,000 international applications have been filed. Most industrialized nations are already parties to the PCT, with more and more countries actively considering adherence to the Treaty (WIPO 1991: 23).

In July 1991, the PCT Assembly adopted a large package of changes (WIPO 1991: 23–24). Some of the more significant changes include: liberalization of the nationality and residence requirements for access to PCT procedures; a system helping applicants avoid omissions and mistakes in designated states in an international application; greater accommodation for filing of computer-generated requests and filing by facsimile machines; simplification of formality and language requirements in relation to fee payments, signature of documents, etc.; modification of the unity of invention requirement toward greater international harmonization of patent laws; improved means for search of certain biotechnology inventions; clarification of international preliminary examination procedures, etc. These changes are designed to streamline procedures for filing and prosecuting an international patent application under the PCT, and to make the use of PCT procedures safer, simpler and more easily accessible to applicants.

Strasbourg Agreement Concerning the International Patent Classification

The Strasbourg Agreement was adopted in 1971 (Leaffer 1990: 531–545). This Agreement encourages unification of national patent legislation and serves to provide a uniform foundation for the search of patents. It is a special agreement under the Union of Paris and any member nation of the Paris Convention may adhere to it. The Agreement requires that member nations apply its international patent classification system, including the symbols of classification, in patents, inventor's certificates and utility models.

The International Patent Classification (IPC), which is administered by WIPO, ascribes a symbol to patent documents which are classified into 114 classes, and over 46,000 groups and subgroups. The adoption of these symbols by the principal patent offices of the world has greatly facilitated international patent searching and the dissemination of the technological information contained in patent specifications. By the beginning of the 1980s a total of over one million patent documents were being published by the world's patent offices annually (WIPO 1983b: paragraph 23).

INTERNATIONAL TRADEMARK AGREEMENTS

Madrid Agreement Concerning the International Registration of Marks

The Madrid Agreement, which was adopted on April 14, 1891 and came into force on July 15, 1892, has been subject to many revisions throughout the past 100 years. It is a special union within the Union of Paris administered by WIPO's International Bureau, and is open only to states adhering to the Paris Convention. It is designed to simplify procedures for persons wishing to secure trademark protection in a number of countries (Leaffer 1990: 229–336).

Where a person has registered a trademark in a country which is a signatory to the Madrid Agreement, that person may, by filing a single application with the International Bureau of WIPO, receive protection for that mark in any other signatory country designated by the applicant. The applicant will enjoy protection for twenty years under the agreement on the same terms as the nationals of the country in which registration is obtained. So far, the Madrid Agreement has been ratified by over thirty countries.

Because the Agreement requires registration in the country of origin as a prerequisite for international deposit, some countries (like the United States and Canada) consider it to favor those countries that have the least demanding requirements for registration, such as a brief examination period. Consequently, countries which have a more elaborate mark examination system are reluctant to join the Agreement. Businesses in the non-member countries may still take advantage of the Madrid Agreement by establishing a subsidiary in one of its member nations and applying for trademark protection through that nation. Interest in adhering to the Agreement is growing in countries such as the United States (Schechter 1991: 419–446).

Trademark Registration Treaty (TRT)

The TRT was concluded at a Vienna diplomatic conference on June 12, 1973, and entered into effect on August 7, 1980 (Leaffer 1990: 291–336). It was designed to establish a truly international trademark. Unlike the Madrid Agreement, the TRT allows for an international registration which does not depend on pre-registration at a national level. A single application can be directly filed by the owner with the International Bureau of WIPO. This has the effect of registering in all signatory countries of the Treaty, on the same terms which apply to nationals of the signatory countries.

As a special agreement under the Union of Paris, only members of the Union may join the TRT. However, the TRT has not been very successful in attracting countries to join it, with only five signatories (mostly African nations) in its first three years of operation. The TRT is designed to avoid the major objection to the Madrid Agreement, which is claimed to favor countries that have lax rules for trademark registration, but it does not really offer much more than the general principles already accepted in the Madrid Agreement (WIPO 1983a).

The Nice and Vienna Agreements

Two international agreements have been established to implement a common system for the classification of trademarks. The Nice Agreement Concerning the International Classification of Goods and Services for the Purposes of the Registration of Marks, concluded on June 5, 1957, created a list of forty-two classes of goods and services and an alphabetical list of goods and services (Leaffer 1990: 499–512). The Nice Agreement came into effect on April 8, 1961, and was revised in Stockholm in 1967 and Geneva in 1977. It has been widely adopted both by signatories to the Agreement and by over thirty countries which apply its classification without having joined the Agreement (WIPO 1981 and WIPO 1987).

The Vienna Agreement Establishing an International Classification of the Figurative Elements of Marks was completed at Vienna in 1973 and entered into effect in 1985 (Leaffer 1990: 546–560). While the Nice Agreement handles the classification of words and names, the Vienna Agreement provides classifications for figurative elements such as images of human beings, animals, plants, landscapes and geometric figures used as trade symbols. The classification includes some 300 divisions and 3,000 sections.

Lisbon Agreement for the Protection of Appellations of Origin

The Lisbon Agreement, which was concluded in 1958, provides for the registration of only those geographic names which indicate a physical quality of goods, such as Limoges (porcelain) (Leaffer 1990: 278–290). The Lisbon Agreement on appellations of origin should be compared with the Madrid Agreement on indications of source. "An indication of source" is merely a geographic designation indicating where the goods came from, e.g. California wines. The primary goal of the Madrid Agreement is to protect the consumer from being misled. But "appellation of origin," which concerns product characteristics or quality arising from the production of goods in a specific geographic location, protects manufacturers of the products concerned.

INTERNATIONAL PROTECTION OF INDUSTRIAL DESIGNS

Hague Agreement Concerning the Deposit of Industrial Designs

The Hague Agreement of 1925 was created to facilitate the international protection of industrial designs (Leaffer 1990: 471–498). Only nations that are members of the Union of Paris may become members of the Hague Union. Protection of an industrial design in other member states can be secured by making a single deposit with the International Bureau of WIPO. The owner of an international deposit enjoys the priority right conferred on members of the Paris Union within the member nations of that Union. An international deposit is protected for a first term of five years and is renewable every five years upon payment of the prescribed fees.

Locarno Agreement Establishing an International Classification for Industrial Designs

The Locarno Agreement, which was signed in October 1968, creates a system of classification for industrial designs for use by the industrial property offices of the world within the Paris Union framework (Leaffer 1990: 513–530; WIPO 1989). The terms of this agreement are virtually identical to those of the Nice Agreement of 1957 on trademark classification. The Locarno Classification comprises 31 classes and 211 sub-classes of design types together with an alphabetical list of some 6,000 categories of goods in respect of which a design may be used. Member countries may use the classification system for whatever purposes the country deems appropriate.

INTERNATIONAL AGREEMENTS CONCERNING BIOTECHNOLOGY

International Convention for the Protection of New Varieties of Plants (UPOV Convention)

The International Union for the Protection of Plant Varieties (UPOV) was established in 1961 as an intergovernmental agreement among a number of European nations to create guidelines for national legislation to provide industrial property protection for breeders of plant varieties (Leaffer 1990: 53–75). UPOV has established minimum standards that each member country must recognize in its national law. UPOV upholds the principle that only one form of protection is to be available for each plant species. Under the UPOV arrangement, a separate application must be filed in each member nation to secure protection in that country. Like the Paris Convention, UPOV allows a breeder who has filed an application in one member nation to enjoy a twelve-month right of priority in other member nations.

Budapest Treaty on the International Recognition of the Deposit of Micro-organisms for the Purpose of Patent Procedure

The Budapest Treaty, which was concluded in 1977 and came into force in 1980, was designed to solve the confusion resulting from the lack of uniformity in national requirements for micro-organism deposits by providing for the single deposit of micro-organism inventions (Leaffer 1990: 125–140). A number of International Depository Authorities (IDAs) are designated to accept deposits. The Treaty requires that all contracting parties recognize the deposit in a single IDA. The Budapest Treaty is a special agreement under the Union of Paris and is administered by the International Bureau of WIPO.

INTERNATIONAL COPYRIGHT AGREEMENTS

The Berne Convention

The most important international agreement on copyrights is the Berne Convention for the Protection of Literary and Artistic Works (Leaffer 1990: 339–394). The Berne Convention was created in 1886, following the calls of several internationally famous authors of the late nineteenth century, such as Victor Hugo and Ivan Turgenev. Nations acceding to the Convention agree to provide certain minimum protection for authors and to meet periodically to discuss new developments, such as changing technology. The Convention has undergone several major revisions for the purpose of expanding the rights of copyright owners or authors. More than eighty countries (including the United States) are now parties to the Convention.

Several provisions are considered central to the Berne Convention (Burger 1988). It requires member nations to give national treatment to authors publishing in any member country. An author who publishes a copyrighted book in Finland will have the same copyright protection in Argentina as any Argentinian author. National treatment relieves the burden of suing infringers in other countries, because the courts only need to use local law to determine the rights of the parties. The Berne Convention also prescribes some minimum copyright standards, such as the duration of copyright protection. Berne Convention members have to provide protection for at least the life of the author plus fifty years for most works, with minimum fifty-year protection for anonymous or pseudonymous works and cinematographical products, and twenty-five-year protection for photographic works and works of applied art. Berne Convention members also agree to remove formalities that restrict copyright. For instance, if a foreign author asserts copyright in the United States, the copyright notice "©" commonly found on most works cannot be required.

One controversial provision that delayed the US accession until 1988 was moral rights or paternity and integrity rights for authors that the Berne Convention requires its members to recognize. The United States historically has not recognized the moral rights of authors on the ground that domestic laws on unfair trade practices provide adequate protection of the moral rights of Berne Convention authors. In the mid-1980s, the United States passed legislation giving minimal protection to the moral rights of authors in limited categories of works.

The Universal Copyright Convention (UCC)

The UCC was formed in 1952 under the auspices of UNESCO as an effort to provide an alternative to Berne to allow the participation of certain countries (such as China, the former Soviet Union and the United States) which originally rejected the Berne Convention (Leaffer 1990: 378–414). The UCC prescribes that the formalities required by the national law of a contracting state shall be considered to be satisfied if all the copies of a work originating in another contracting state carry the symbol "©" accompanied by the name of the copyright owner and the year of first publication (Kerever 1991: 50).

Under the UCC, member nations must protect a copyright for at least the life of the author plus twenty-five years. However, nations that already had more restrictive terms for certain classes of works could limit the copyright terms to twenty-five years from the date of first publication. The UCC also requires its member nations to extend national treatment to the published works of nationals from other signatories. Countries that are members of both the Berne Convention and the Universal Copyright Convention must comply with the terms of the Berne Convention wherever the two treaties conflict. Currently, more than eighty countries are parties to the Universal Copyright Convention.

THE TRADE RELATED ASPECTS OF INTELLECTUAL PROPERTY RIGHTS (TRIPS)

With the growth of world trade in the 1980s, international differences in protecting intellectual property have caused an increasing number of trade disputes. The Uruguay Round negotiations symbolized a watershed in international trade policy by including intellectual property rights on the agenda. The TRIPs Agreement of 1994 establishes much higher standards of protection for a full range of intellectual property rights than are embodied in any existing international agreements, and provides for the effective enforcement of those standards both internally and at the border (Hill 1994: 10–11). The intellectual property rights covered by the Agreement include: copyrights; patents; trademarks; industrial designs; trade secrets (undisclosed information); integrated circuits (semiconductor chip mask works); and geographical indications.

According to the TRIPs, all countries are given one year to implement the Agreement. Developing countries are allowed an additional four years for implementation, except for provisions concerning patent protection for pharmaceutical and agricultural products, where they are given an additional nine years. Developing countries must, however, at the end of the one-year transition period granted to all countries, provide national treatment and most-favored-nation treatment, while allowing limited exceptions thereto for agreements administered by the World Intellectual Property Organization. Acting as the dispute settlement body, the General TRIPs Council will handle disputes involving the Agreement.

THE WORLD INTELLECTUAL PROPERTY ORGANIZATION (WIPO)

WIPO, established by a convention signed at Stockholm on July 14, 1967, became a specialized agency of the United Nations in 1974 (Leaffer 1990: 563–581). The origins of the organization can be traced back to the Paris Convention of 1883 and the Berne Convention for the Protection of Literary and Artistic Works of 1886. The two conventions provided for the establishment of two separate international secretariats, which were staffed by officials of the Swiss Government. In 1893 these two offices were merged under the name of BIRPI, the acronym of the French language version of its name, the United International Bureaux for the Protection of Intellectual Property.

WIPO has traditionally operated three kinds of activities: registration, the promotion of intergovernmental cooperation in the administration of intellectual property and specialized programs. Via these activities, WIPO has provided valuable assistance to developing countries in terms of technology transfer. The registration activities of WIPO include the administrative duties conferred on it by various international, industrial and intellectual property conventions. For instance, it is the registration agency designated by the Patent Cooperation Treaty and the Trademark Registration Treaty. By carrying out these registration duties, WIPO provides industrial property services to developing countries.

In terms of promoting intergovernmental cooperation in the administration of intellectual property, WIPO launched its Program for Patent Information and Documentation in 1975, through which it provides, gratuitously, search reports to organizations in developing countries. These search reports are provided at the expense and from the resources of some participating developed countries. WIPO has also been involved in the establishment of a patent information and document center (CADIB) within the framework of the African Intellectual Property Organization (OAPI) to serve the needs of the French-speaking African countries, and a similar information and document center (ESAPADIC) within the framework of the Industrial Property Organization for English-Speaking Africa (ESARIPO).

WIPO also transfers technology to developing countries through various development assistance programs. This assistance is reflected in three areas: first, WIPO provides educational assistance to train government officials from developing countries on the basic concept and protective system of intellectual property; second, it helps developing countries in promulgating their appropriate legislation, including the formulation of various model laws for developing countries on inventions and industrial designs, marks, trade names, licensing practice, etc.; third, it assists developing countries in establishing appropriate administrative infrastructures (Blakeney 1989: 25).

There are currently a number of ongoing discussions of WIPO that may have an impact on the future development of international intellectual property protection. These include a patent law harmonization treaty, a trademark law harmonization treaty and a possible protocol to the Berne Convention. In response to the criticism that the weakness of WIPO-administered treaties lies in their lack of dispute resolution mechanisms, the WIPO Secretariat has prepared a draft treaty, which calls for the use of consultation, good offices, conciliation and mediation for the resolution of disputes (Gorlin 1993: 179–182). One direct result of these discussions is the establishment of the WIPO Arbitration Center in July 1994 (see Chapter 16).

SUMMARY AND CONCLUSION

This chapter has reviewed several of the most important international conventions, treaties and agreements. Together, they lay down a significant foundation for the protection of intellectual property. While these agreements do not preempt national laws, they establish guidelines that facilitate technology transfer throughout the world. In the past decade, the issue of intellectual property protection has become increasingly important in global trade. The conclusion of the TRIPs in December 1994 has provided a tremendous boost to the international efforts to protect intellectual property, as it establishes considerably higher standards of protection for a wide range of intellectual property rights than were found in existing international agreements. The area that needs the most improvement seems to be the international protection of trade secrets, where currently there is no effective international system.

Meanwhile, multilateral cooperation on the protection of intellectual property has made substantial progress on a regional basis. For instance, fourteen European nations, through the Convention on the Grant of European Patents, have established a European Patent Office that issues patents recognized in member states. All of the European Union nations (except Ireland and Portugal) have signed the Convention. Canada, Mexico and the United States, through the North American Free Trade Agreement (NAFTA) negotiations, have made substantial progress in harmonizing intellectual property regulations within North America. The Agreement prohibits discriminatory treatment in the regulation of intellectual property and is also the only international treaty that extends protection to trade secrets.

7 The issue of intellectual property protection in developing countries

INTRODUCTION

Developing countries have often been accused of inefficiency in protecting the property rights of knowledge, tolerating instead piracy practices. Developed countries' reactions to such problems usually include threats and commercial retaliatory measures. For the fully industrialized countries, intellectual property rights (IPRs), such as patents, provide compensation for risks incurred in the innovative process, and therefore function as an incentive for such research and development (R&D) activities. The company obtains its reward for technological leadership through the monopoly provided by the patent system (Mansfield 1990: 17–30).

Developing countries on the other hand have a very different point of view on the issue. Innovation there progresses along a path different from the developed countries, often beginning with a product's introduction and ending with control of the technology. The dynamics and standpoint of protecting innovative activities are entirely different. For many developing countries, the central concern of government has been with their own development to meet the basic needs of a rapidly growing population (Vaitsos 1972: 71–97). IPRs, like patents, are often viewed as agents that build up the monopoly of their economies and technological development by foreign multinational corporations (MNCs). This poses difficult barriers for the development of indigenous firms.

These different perspectives have given rise to major gaps in the protection of IPRs between the developed and developing countries. This chapter reviews the traditional views of developing countries on IPRs, examines the intellectual property system of Brazil and discusses the costs and benefits of IPR protection for developing countries. The future trends of development in the developing countries is also considered.

VIEWS OF DEVELOPING COUNTRIES ON IPR PROTECTION

For a relatively long period after the end of the Second World War, economists held fairly negative attitudes about the benefits developing nations might expect to extract from IPRs. The prevalent approach embodied in the pre-1970 literature is well represented in the statement by Penrose (1951: 220): ". . .non-industrialized countries and countries in the early stages of industrialization gain nothing from granting foreign patents since they themselves do little, if any patenting abroad. These countries receive nothing for the price they pay for the use of foreign inventions or for the monopoly they grant to foreign patentees." She concluded her analysis by suggesting that developing countries "should be exempt from any international patent arrangement."

During the 1970s, this negative view toward intellectual property protection in developing countries grew even stronger along with a general trend in favor of regulating technology transfers and foreign capital in developing countries. There was a rather heated debate over technology licensing arrangements under which technology is transferred from a MNC to a local firm or its affiliate. The literature emphasized that developing countries paid an unfairly high price for such transfer. Too much rent was extracted in the form of profits, royalties, transfer pricing advantageous to the supplying firm and management or training arrangements (UNCTAD 1987; and Rapp and Rosek 1990). To control the perceived problems, a number of developing countries enacted technology-transfer legislation or joined a treaty of this nature, such as the Andean Pact. Some even suggested that developing countries should abandon the patent system.

This development also fueled the North–South debate on the so-called New International Economic Order (NIEO), with developing countries seeking the establishment of an International Code of Conduct on the Transfer of Technology (UNCTAD 1985). This code, intended to affirm the right of nations to review technology-transfer contracts, could make it far more difficult for MNCs to impose various commonly used restrictive clauses. In parallel, there was also growing pressure by developing countries for radical revision of the Paris Convention. The thrust of the proposed reform was to loosen the international standards of industrial property protection. However, the developing countries were not successful in their push for reform, because of the opposition of developed countries, particularly the United States (Kunz-Hallstein 1989: 269).

The following is a summary made by Robinson (1988: 145–148) of the traditional views of developing countries on patents, trademarks, copyrights and trade secrets. It should be borne in mind that some of these views have already undergone major changes in many developing countries in the last decade.

Many developing countries find it difficult to accept the traditional notion of a patent due to a number of considerations. First, foreign-owned patents should be treated differently from locally developed patents with regard to period of protection and compulsory licensing requirements on the following grounds:

1 As R&D allocation in the developed countries is not influenced by patent protection in the developing countries, such protection does not have direct impact on the innovation in the developed countries.
2 Most patents issued by the developing countries are never used in those countries, with many of them being used simply to exclude competitors from accessing the markets of developing countries. Some patentees, for example, may take advantage of the right to exclude imports of a non-licensed manufacturer in regard to a patent protected locally, even though no local production is carried out.

Second, when patents of foreign owners are not used, local firms have to wait for a long period to be able to use them. Under present legal systems, even where compulsory licensing is indeed enforced, it normally takes a long time to compel licensing. There is a one-year priority for local filing, plus three years during which the compulsory licensing rule is not yet applicable, another two to three years to establish liability for not utilizing the patent, plus perhaps another two years before a local court issues orders of compulsory licensing. By that time, half of the useful life of the patent has passed by, and the technology may already be out-dated.

Third, no criteria have so far been tested with regard to how patent protection contributes to a country's economic development. In some cases, public welfare may be badly served by issuing patents to cover such essentials as pharmaceutical and food products. A foreign parent corporation can license a patent to a local firm over which it maintains control, in which case the inappropriate extraction of royalties is viewed by the developing countries simply as an instrument to drive down local profits as well as local tax liability.

Trademarks represent a different problem. The use of an internationally known trademark or name by a firm of a developing country may be highly beneficial in gaining access to both the local and international markets. However, if the firm is restricted to using the licensed foreign mark or name, it becomes captive to the foreign firm owning that mark or name. Subsequently, the local firm cannot develop any mark or name of its own. To solve this problem, the Mexican government has required that any Mexican firm licensing a foreign mark or name should also use its own mark and name in an equally active way, whether on products for local consumption or for export. By doing this, the local firm can gradually become independent of the foreign mark or name and use its own. Another problem is that the local licensee in a developing country may not be allowed by the licensor to use its local name or mark in external markets for fear that the licensee will not maintain product quality. The non-use of the foreign mark or name hardly makes it possible for the developing country firm to compete on foreign markets.

Copyrights represent another problem. It has been considered by many to be in the interest of developing countries to copy and translate as many foreign books and journals, films, tapes and as much computer software as possible without paying royalties. As transfer of technology in the public domain is thereby expedited, many have thought this to be a highly recommendable option for poor developing countries. Moreover, owing to lower levels of education, developing countries normally produce far fewer works than developed countries for which they would like foreign copyright protection. Therefore, there is really no reciprocity for purchasing these works. For many, stealing these kinds of products from rich countries and foreign corporations is not considered immoral, as local firms are driven by a sense of mission to obtain foreign technology as cheaply as possible in order to facilitate local economic and technological development.

The transfer of technology into developing countries via licensing based on secrecy also presents a serious problem. Many argue that it is not really in the interest of the receiving country to render that secrecy enforceable. One of a developing country's goals in obtaining foreign technology is to rapidly disseminate the technology throughout its society, thereby promoting more effective modernization of productive activity. Moreover, the legal systems in many developing countries are not helpful for the efficient enforcement of confidential provisions in employment contracts or for non-disclosure clauses in technology transfer agreements.

THE SYSTEM OF INTELLECTUAL PROPERTY PROTECTION IN BRAZIL

Brazil's trade pattern demonstrates the typical pattern for a developing country. In most patent-intensive goods, Brazil has apparent disadvantages due to a limited technological base in comparison to the industrial countries. But in some goods subject to trademark protection, such as luggage, clothing and furniture, it maintains net-export positions. For many Brazilian companies, there is a clear advantage in having laws that weakly protect the intellectual property of foreign competitors, allowing local firms to imitate foreign technologies. As Brazil grows technologically more sophisticated, stronger laws may become necessary (Marcus 1993: 11–25).

Brazil was one of the original adherents to the Paris Convention, though it has not signed the provisions of the Stockholm text. The Brazilian 1971 Industrial Property Code describes patent and trademark regimes, and marginally covers the trade secrets area. Its 1973 Copyright Law provides for copyright protection. The areas specified by the Industrial Property Code are under the administration of the National Industrial Property Institute (INPI). INPI's mandate comprises registration, protection and assignment of patents and trademarks. Policy issues pertaining to copyright protection are handled by the National Copyright Council (CNDA), while copyright registration is decentralized to eleven institutions.

Patent protection

Brazil follows international practice: for a standard invention patent to be granted, an invention must be "novel," "non-obvious" and "commercially useful." All patent applications should be filed with INPI. For foreign patents (those filed first in a foreign country with which Brazil maintains an international treaty for mutual protection), applications must be filed in Brazil within the "period of priority." The Paris Convention requires one year for patents and six months for models and designs. If not filed in Brazil within this period, the invention automatically falls into the public domain. Standard invention patents in Brazil are protected for fifteen years from the date of filing whereas other types of patents (utility models, industrial models and designs) are protected for ten years, but the overall time of protection is much less because the period of processing from filing to actual grant averages four years (Turner 1988: 16).

In a few sectors, patents either are not granted or are restricted. Many of these were initially included in the 1945 Patent Law, which did not permit product patenting in pharmaceuticals, chemicals and foodstuffs, though Brazil did recognize patents for processes. The law conformed to the then prevalent international practice of denying patent privileges in areas of great social impact, where the major governmental concern was ensuring an adequate domestic supply. Brazil's restrictions on pharmaceutical and other products were expanded in a 1969 law, and then written into the Industrial Property Code. Pharmaceuticals is probably the non-patentable product group which has caused the most frequent intellectual property rights friction with developed countries (Frischtak 1990: 67).

The main reason for Brazil stopping recognition of patents for pharmaceutical processes by the end of the 1960s was possibly the ability of native firms to produce drugs. By the late 1980s, fifty MNCs still dominated 80 per cent of the Brazilian market. Four hundred and fifty native companies focusing on several therapeutic groups where few companies competed shared in the group of ethical products. Native companies shared in 10 per cent of total domestic production, indicating a developing industry characterized by foreign dependency. The international trade balance of the sector remained negative, constituting 20 per cent of the sector's total sales (Vasconcellos and Pereira 1994: 234). Meanwhile, Brazil's statutory prohibitions against product and process protection for pharmaceutical patents led to an action against them in June 1987 under section 301 of the US Trade Act of 1974.

Other major non-patentable sectors include chemicals, metal alloys and mixtures, and atomic substances and materials (Frischtak 1990: 67–69). For chemicals, the Code precludes patents on substances, matter or products obtained by chemical means or processes, though the processes for obtaining or transforming such substances can receive patents. The patentability of agrochemicals (and other fine chemicals) is not described at all. In addition, the Code has not defined two important growing areas: integrated circuit design and biotechnology products and processes.

There is a widespread criticism that the system of patent protection in Brazil is insufficient with its limited coverage and strict working requirements. In accordance with its legislation, once a patent is awarded, owners should work the patent within three years and without interruption for more than a year. Otherwise a patent holder may be forced into compulsory licensing to a third party, who can request from INPI special, non-exclusive rights to exploit the patent. Compulsory licensing can also be imposed if the use of the patent cannot meet market demand.

Trademark protection

Legislation on trademarks is generally in line with international standards. Registration of a mark, advertising expression or device is awarded for ten years from the date of issue and is renewable for identical and successive terms. A renewal can be obtained by filing an application during the last year of each ten-year period. Trademark registration activity in Brazil is quite intense (Frischtak 1990: 70–71), but trademark protection is diluted by poor enforcement. There are also regulations designed to favor local registrants at the expense of foreign applicants.

One notable example is registrations on a "first-to-file" basis, where INPI encourages abuse by allowing Brazilian firms to register foreign trademarks as their own, provided they have not already been registered locally by the foreign trademark holder. Even internationally recognized marks can be appropriated by third parties if they are not registered in Brazil on a timely basis. Moreover, registration privileges may lapse on petition from any interested party if the mark is not launched within two years from the time of registration, or if use is discontinued in Brazil for more than two consecutive years, though *force majeure* excuses non-use against a petition for trademark cancellation. While the courts may accept the defense that non-use resulted from trade restrictions, INPI does not. There are many cases in which these provisions have been abused by local firms in bad faith (Turner and MacNamara 1989: 15).

Copyright protection

The 1973 Copyright Law provides protection for written works, architectural and engineering designs and plans, as well as film, photographs, music video and other forms of creation. Copyrights are good for sixty years beyond the life of the author, or for recordings and broadcasts, sixty years after their creation. Responsibilities for collection and registration of works for copyright are distributed among ten specialized institutions ranging from the National Library (which accepts book registrations) and the National Cinema Council (CONCINE) to the Federal Council of Engineering, Architecture and

Agronomy, which registers designs and plans. While the Copyright Law does not require registration as a prerequisite for obtaining protection, it is recommended to register anyway as a deterrent to piracy.

In 1987, the government passed legislation providing a *sui generis* form of copyright protection for computer software, rather than recognize protection under its existing copyright statute. Software was considered more a utilitarian work than a literary or artistic one. In December 1988, CNDA appointed INPI as the agency responsible for registering computer programs for copyright. Registration normally takes fifteen to thirty days and is valid for five years. It is renewable for twenty-five years from the date the software is marketed. To be sold domestically, both national and foreign software needs to be cataloged with the Special Secretariat of Information (SEI) as a condition precedent to commercialization and distribution.

SEI's acceptance of software depends on a "similarity" test. While such tests have not been a major obstacle to imports of weakly differentiated software, SEI could ask for disclosure of computer source codes by requesting that specific software be subject to an INPI review. This may subsequently lead to an INPI-mandated transfer agreement, with disclosure of source codes. This is considered inconsistent with traditional copyright doctrine. Under the Berne Convention and the Universal Copyright Convention, an author would not have to describe his work or the document used to develop it.

In terms of the videocassette industry, CONCINE Resolution 136 of 1987 prohibits the sale, rental and exchange of videocassettes unless they bear an authorized CONCINE stamp. The Resolution also makes it known that video clubs violating this prohibition may be raided or closed by forfeiting their status as non-profit organizations. But lax enforcement continually costs businesses in the industry hundreds of thousands of dollars. It has been acknowledged by the local film industry that approximately 70 per cent of all videocassette titles marketed in Brazil (a $160 million market) are illegal copies. CONCINE only enforces Resolution 136 against videocassette titles whose entry CONCINE had sanctioned. Resolution 136 is not applied to titles entering Brazil illegally (Turner 1988: 16).

Trade secrets protection

Trade secret is still a relatively misunderstood concept. Brazil provides very weak statutory protection against disclosure of trade secrets to competitors, because the country does not have a statute that specifically provides protection for trade secrets or know-how. There is only an indirect form of protection, usually of five years' duration, derived from the unfair competition section of Brazil's Industrial Property Code, Article 178, Item xii, which states: "an unfair competition crime is committed by whoever discloses or exploits, without authorization, when performing services for others, a manufacturing secret, which was entrusted or which he came to have knowledge of as a result of his work." The statute also can be applied when the unauthorized act is performed after an individual has left his former employment (Turner 1988: 15).

However, if a company obtains trade secrets by hiring away employees, it normally cannot be sued. Companies are considered liable only if they are found to have acquired trade secrets by resorting to "unfair means," such as through industrial espionage. The burden of protection rests mainly with companies which are potential targets of infringement. As a common practice, Brazilian courts have dismissed most cases of trade secret infringement on the ground that plaintiffs failed to take effective measures to protect their trade secrets. Lack of adequate self-protection may also lead the courts to the conclusion that information of insufficient value has been inappropriately classed as a trade secret (Richards 1988: 180).

Technology transfer

Brazil's technology transfer regulations influence most IPR protection. By law, INPI requires that all technology transfer agreements and all patent and trademark licenses must be approved and registered with it, if they are to be enforceable. INPI's approval policy is based on the assumption that tight control is necessary to minimize foreign exchange outflows and to develop Brazil's base of advanced technology. Consequently, INPI treats technology-transfer agreements more as instalment contracts for sale of technology than as licenses. Payments for the technology are also regarded as instalment payments rather than royalties.

INPI also believes that the technology transferred should belong to Brazil within the shortest possible period and that very few Brazilian licensees have adequate economic leverage to negotiate a fair agreement. Therefore, it intervenes vigorously on the side of Brazilian companies. INPI usually has approved agreements where Brazil has a developmental need for the technology, where the transferred technology may help promote Brazilian exports or where there is no similar technology in Brazil. The normal duration of technology transfer agreements that INPI permits is five years, though it is possible to renew for an additional five years (Turner 1988: 14).

Assessment

Overall, protection of intellectual property in Brazil is relatively weak. Although protection is granted in some areas, important gaps exist in the legislation on patenting, copyright, trademark and particularly trade secrets. Some of these gaps are the products of broad development policy decisions: for example, the lack of patent protection for pharmaceutical and food products resulted from public health concerns, which are quite common among developing countries. In a number of other instances, however, these gaps can be traced to inherent weaknesses in the IPR administrative system – INPI's slow processing of patent applications is a case in point. INPI's approval policy for technology transfer projects unfavorably affects most IPR protection. Finally, time-consuming legal processes, small penalties for infringing existing legislation, and limited enforcement capabilities have rendered the system of IPR protection in Brazil very inefficient.

The last problem has been particularly prominent. The Brazilian judicial system does not seem to provide an effective deterrent to violations of intellectual property rights. Proceedings are widely criticized as being too slow. Civil and criminal penalties for violating intellectual property rights in Brazil are small compared to those set out in the legislation of most developed countries and handed down in court decisions. In Brazil, intellectual property rights violations have not brought about public pressure for swift and effective action. This probably mirrors the prevalent perception of the Brazilian public that there is not much to gain in strengthening the system of intellectual property protection (Frischtak 1990: 73).

In the case of patent infringement, violators are theoretically subject to civil and criminal penalties. But there has been a reluctance to throw an infringer into prison or to order sufficient compensation for damages. There are neither preliminary nor permanent injunctions. As a civil law country, Brazil allows very limited use of case precedent (*stare decisis*), resulting in contradictory court decisions on similar sets of facts. Judges would also normally not authorize inspections of an alleged infringer's factory, without a clear case for infringement. In fact, it is rarely possible to establish process infringement without such an inspection.

Criminal penalties for trademark violations may lead to three months' to one year's imprisonment, in addition to assessment of damages. But the law has not been very effective in preventing abuses of the "first-to-file" rule. Copyright infringement may be also subject to civil and criminal penalties. Unauthorized copies can be seized under court order, while the violators are required to reimburse the rights holders with all the income from illegal sales. Furthermore, moderate monetary fines can be imposed on violators. However, the only penalty that is considered to have some deterrent impact is imprisonment, with terms ranging from three months to four years. Penalties for violation of software property rights can result in fines and detention for as few as six unauthorized copies of the software. Lax enforcement, however, remains a serious problem (Turner 1988: 16).

RETHINKING INTELLECTUAL PROPERTY PROTECTION IN DEVELOPING COUNTRIES

By the mid-1980s, developed countries regained ground in the international debate on intellectual property rights. Technological developments have greatly raised the economic value of knowledge appropriation as the world economy becomes more R&D intensive, copying and imitating become easier, economic globalization continues, and traditional jurisprudence on intellectual property rights is challenged. It is not surprising to see an increasing concern among developed countries with intellectual property. The use of trade laws by the United States and the European Community in their fight against "piracy," and the progress in the discussions on TRIPs in multilateral trade negotiations, played an important role in this process.

During the same period, the approach of developing countries toward foreign direct investment and technology transfer has undergone major changes as a result of a number of changes in the course of their development and international environments. The foreign debt crisis, counter-productive experiences of an inward-looking development strategy and highly regulatory policies, decreasing capital flows and economic stagnation and hardships in many developing countries and former socialist countries are some of the reasons for the change of attitudes. There was also a renewed interest in assessing the potential benefits of improving protection of intellectual property among developing countries, though no specific conclusion has been reached. The most notable changes of attitude are found in the newly industrializing countries, which have begun to feel the necessity to protect their own intellectual property in other developing countries. This is evidenced by the increasing frequency of cases involving companies from former piracy countries, such as Taiwan and South Korea, against infringement of their intellectual property in countries like China and Indonesia.

During the past few years, bilateral disputes and negotiations under the threat of unilateral trade sanctions have become the norm. Both the United States and the European Community have expanded their laws and developed commercial policies to handle violations of intellectual property rights. Trade sanctions have been imposed on some countries. In October 1988, the United States imposed 100 per cent punitive tariffs against $39 million of Brazilian goods as a result of an investigation (under section 301 of the Trade Act 1974) of Brazil's refusal to extend patent protection to pharmaceutical products. The investigation was motivated by complaints from the US Pharmaceutical Manufacturers' Association. In December 1987, the European Community suspended its Generalized System of Preferences (GSP) benefits for Korean products, because Korea favored the intellectual property rights of US nationals under the terms of a section 301 agreement (Brueckmann 1990: 305).

These external pressures, under the new environment, have become very effective in bringing about reforms in several developing countries. Some have significantly revised their technology-transfer laws or enforced them less strictly. For instance, the Andean Pact (signed by Bolivia, Colombia, Ecuador, Peru and Venezuela) modified some of its restrictions on direct investment (Decision 220 of May 11, 1987), and Mexico announced new regulations in January 1990 that dramatically relaxed that country's 1972 and 1982 legislative restrictions on technology transfer, patents and trademarks. Major changes have been made in the intellectual property regimes of Korea, Taiwan and Singapore. In January 1992, China was pressured into signing a Memorandum of Understanding with the United States to improve its intellectual property protection.

Meanwhile, there is a renewed interest in the developing countries in the reassessment of the costs and benefits of a good intellectual property system, though there is a difference of views on this point. Among the most comprehensive recent studies is one made by Carlos Primo Braga for the World Bank (1990: 69–87). This study reviews the relevant literature and closely examines the major costs of concern in developing countries and the major benefits that many in developed countries have been arguing for. Primo Braga has made the following discoveries.

The actual system cost, including the cost of introducing and maintaining intellectual property, is in fact not unreasonably high, particularly when a country adopts registration rather than a full-fledged examination system, and introduces its enforcement mechanism in a gradual manner. There are also many ways to improve the effectiveness of these IPR systems without incurring a major financial burden. Incremental royalty payments are much less of a cost in the balance of payments than some of the estimates discussed in public debate, particularly as compared to overall expenditure on imports of disembodied services. The risk of anticompetitive behavior by the intellectual property owner, which is widely considered to result in higher prices and higher entry barriers to newcomers, can be partly contained or reduced. One cost clearly identified is derived from the diversion of resources from certain economic sectors by new R&D activities as a result of stronger intellectual property protection. Empirical tests still need to be conducted on the estimated economic losses resulting from setting up and operating mechanisms to control or eradicate pirating activities.

Primo Braga has also found that the benefits of protection are as difficult to establish as the costs. Most of the benefits that have been projected to grow from stronger protection in developing countries have not been empirically ascertained. These include the favorable influence on domestic R&D activity, the knowledge diffusing impact of patent disclosure and its positive impact on technology transfer through licensing, the push to greater world technological cooperation resulting from better protection in all countries, and beneficial effects on capital formation; particularly through more foreign direct investment. In fact, the protection of IPRs is considered by many governments to be neither necessary nor adequate for strong technology activity. Taiwan and South Korea are often quoted as good examples of countries that successfully advanced their technological development before reforming their IPR system.

Therefore, he concludes that:

> the only clear-cut conclusion – one almost too obvious to state yet no more quantifiable than preceding ones – concerns the risk of trade retaliation, which for countries following an outward-oriented development strategy appears relevant enough to justify a review of the international acceptability of their intellectual property regimes.
>
> (Primo Braga 1990: 87).

This conclusion has been abundantly evidenced by what has happened to Taiwan, South Korea, Singapore, Thailand, Brazil and most recently China. It seems so far that trade policy has been the most efficient way to force improvement of intellectual property protection system in many developing countries.

SUMMARY AND CONCLUSION

This chapter has reviewed traditional views of developing countries on the protection of intellectual property. From a narrow development point of view, many of these arguments are not without some basis. Nevertheless, these traditional views had tremendously negative impacts on the development of the international intellectual property protection system from the 1950s to the early 1980s. A review of the Brazilian intellectual property protection system assessed both its strengths and weaknesses. Many of those weaknesses are associated with the traditional views discussed in the first section. Finally, development in the 1980s is reviewed and the costs and benefits of enhancing intellectual protection are discussed.

The overall conclusion is that improvement of intellectual protection at the global level has become a trend, whether developing countries like it or not. Changing patterns of trade and the increasing importance of intellectual property to the exports of developed countries have been the most important catalysts for the change. The rise of the newly industrialized countries has given them more motivation to improve their legal mechanisms for protection of intellectual property. Changing trade policies of developed countries in the 1980s have played a key role in pushing the developing countries to change their traditional views and practices on the protection of intellectual property. The conclusion of the Trade Related Aspects of Intellectual Property Rights (TRIPs) Agreement at the end of 1994 has raised hope among many that a truly international enforcement system, acceptable to both developed and developing countries, can eventually be established.

8 Enforcing intellectual property protection

INTRODUCTION

Once a company has obtained intellectual property protection, it then faces the far more difficult challenge of enforcing those property rights in the worldwide marketplace. The efficiency of enforcement depends on many factors, such as local laws, the attitude of local officials and the resourcefulness of the intellectual property owner. Many countries consider intellectual property as a private right to be enforced by its owner. Enforcement must be achieved through local law. In many countries (including the United States), intellectual property rights (IPRs) are enforced by a civil suit against infringement and the intellectual property owner may be awarded damages or an injunction against such infringement.

The most serious problem for the intellectual property owner is one of counterfeit or pirated products. Businesses lose billions of dollars each year to commercial counterfeiters. While counterfeiting exists in all industries, there has been an explosion of piracy of video/audio recordings and computer software in the past few years. For example, worldwide, the cost of piracy of software products is estimated to be between $10 and $20 billion in lost revenue (*Financial Times*, December 10, 1992: 7). In recent years, the gray market issue has drawn increasing attention, as a result of the rapid growth of international trade and international licensing of trademarks. Both governments and individual businesses have an interest in stopping piracy and control some gray market practices. This chapter explores several ways of enforcing IPRs against unfair competition from foreign pirates and unfair gray market practices.

BUSINESS RESPONSES TO COUNTERFEIT GOODS

Anyone who has been offered phoney Levi's jeans, pirated TDK cassette tapes or counterfeit Nike sneakers can fully appreciate the difficulty in enforcing IPRs. Counterfeit goods often resemble the genuine product in appearance and quality to consumers. By doing this, counterfeiters take a free ride on the success of the real product.

Intellectual property piracy has infringed upon the interests of the legitimate intellectual property owners in several ways. First, it deprives the owner of his or her hard-earned revenue from the creation and development of the product, because the counterfeiter pays no licensing fees. Second, when the quality of the counterfeit product is substandard, consumers who believed that they bought the genuine product will have a wrong impression of the company that owns the IPRs; the company may suffer damages in image and sales. Finally, the pirated sales steal from the sales of the legitimate dealers, who have a special relationship with the rights' owner and have spent money on sales promotion.

Therefore, a strategy for protection of IPRs at home and abroad is crucially important for any initiative to export overseas. Adopting appropriate measures to ensure protection of IPRs is essential for a number of reasons (Smith 1988: 14): it is a useful instrument for one to keep or regain a competitive edge over one's business rivals; it helps deter counterfeiters from manufacturing the product of the property rights owner; it contributes to continued R&D endeavors by preventing other companies from reproducing those endeavors at minimum cost; it is conducive to goodwill between producer and consumer since it helps decrease consumer purchases of inferior bootleg products; and it serves to guarantee the continued existence of an important source of international royalty income.

When a property rights owner finds someone selling counterfeit goods in any country where the owner has IPRs, he is entitled to take an appropriate legal action for copyright, patent or trademark infringement. Generally, most countries authorize Customs officials to seize infringing goods upon import. US law includes several typical provisions allowing Customs to stop infringing products at the border. Section 526 of the Tariff Act of 1930 blocks imports of goods bearing a US registered trademark without authorization from the trademark owner, and also authorizes Customs to seize those goods. Section 337 of the Tariff Act of 1930 provides similar protection from imports that infringe upon US patents. Section 602 of the

Copyright Act prohibits the import of products that infringe on US copyrights, and authorizes Customs to confiscate any such products.

In the United States, the owner may obtain protection against importation of infringing goods by recording a trademark or copyright with the US Customs Service. In 1988, Congress strengthened the means available for preventing import of infringing goods. Using section 337 of the Tariff Act of 1930, any owner of a registered US intellectual property right who is convinced that an import infringes on that right may apply to the International Trade Commission (ITC) for relief (Mutti 1993: 339–357). The ITC has the power to issue orders excluding goods from the United States, ordering unfair trade practices to cease, and in some instances ordering seizure of the offending goods. Although section 337 was determined to be in violation of GATT in 1989, it is viewed in the United States as playing a key role in protecting IPRs. The following example is used to illustrate the role of the ITC.

Hyundai Electronics Industries, Ltd v. United States International Trade Commission

899 F.2d 1204 (1990)

Federal Circuit Court of Appeals

Facts Hyundai entered into an agreement in South Korea to produce erasable programmable read only memory chips (EPROMs) for the General Instrument Corporation. The agreement required Hyundai to produce the EPROM chips to General Instrument's specifications, but allowed Hyundai to use excess chips for its own products. General Instrument took possession of the chips in South Korea, flew them to Taiwan for further processing, then imported some of the chips into the United States.

Intel Corporation, a US business, alleged that the EPROMs infringed four of its patents, and filed a complaint under section 337 with the ITC. The ITC found that the chips did infringe Intel's patents. It ordered the exclusion of all EPROMs manufactured by Hyundai to General Instrument's specifications, whether imported by themselves or incorporated into circuit boards or other carriers. The order further excluded all Hyundai products that used EPROMs, including computers, computer peripherals, telecommunications equipment, and automotive electronic equipment unless Hyundai certified for each shipment that it had made "appropriate inquiries" and determined that the goods imported in the shipment did not contain EPROMs covered by the exclusion order.

Hyundai appealed the order of the ITC, claiming that, by including secondary products, the relief granted was far too broad.

Issue Did the ITC exceed the scope of its authority by ordering certification of secondary products?

Decision No. The decision of the ITC is affirmed. The Commission's limited exclusion order requiring Hyundai to certify, as a condition of entry, that certain of its downstream products do not contain infringing EPROMs is a reasonable accommodation. . . . The Commission found that Hyundai had violated section 337; that specific EPROM chips embodied the violation; that Hyundai remained free under its manufacturing agreement with General Instrument to use excess infringing EPROMs for its own requirements; and that Hyundai could easily assemble the infringing EPROMs into and import them as part of other Hyundai product "containers" that require EPROMs to function, including wafers, circuit boards, computer, computer peripherals, telecommunications equipment, and automotive electronic equipment. It concluded that the certification provision "is a reasonable means of ensuring the effectiveness of the remedy to which Intel has proven itself entitled. . . ."

(Hotchkiss 1994: 313–314, reprinted by permission of McGraw-Hill, Inc.)

The US Customs Service has also recently concentrated on growing concerns over the import of counterfeit goods. While celebrating its bicentennial in 1989, Customs announced that in its third century, it will rechannel its commercial enforcement thrusts toward the protection of technology, rather than merely continuing in its more traditional role for protection of revenue. Importers are advised to pay due attention to this shift in emphasis by Customs. In 1989 alone, Customs created both a new Office of Trade Initiatives and an Intellectual Property Rights Task Force at its Washington DC headquarters. In addition, Customs issued proposed regulations on the protection of semi-conductor chips and launched a new trade initiative program, known as "Project Cicero," to seek the cooperation of US industry, labor and consumers in investigating intellectual property rights violations (Baker 1990: 38).

Since its passage, the Trademark Counterfeiting Act of 1984 has become increasingly important in the fight against counterfeiting.

Managers who take adequate precautions and resort to the provisions of the law can help stem counterfeiting traffic. Under the Anti-Counterfeiting Act, the trademark owner may go to federal court and obtain a temporary restraining order (TRO) as well as a search and seizure order against the infringer. As soon as the TRO is granted, the trademark owner should lose no time in coordinating with the federal marshal service to execute the TRO. Once the search and seizure order has been implemented, with counterfeit merchandise having been found and seized, the plaintiff may be able to obtain a preliminary and then a permanent injunction (Brooks and Gellman 1993: 49–51). The Trademark Counterfeiting Act of 1984 imposes heavy criminal penalties, which can amount to a fine of as much as $1 million and fifteen years' imprisonment of individuals (PTO 1991: 7).

Piracy of copyrighted materials may also be subject to criminal penalties. In the United States, a person who wilfully infringes a copyright for financial gain is subject to a $25,000 fine, one year imprisonment, or both. If the offense involves a substantial number of infringing copies of phonorecords or motion pictures or trafficking in counterfeit labels for phonorecords, motion pictures or other audio-visual works, the penalties may be as much as $250,000 and five years' imprisonment. Furthermore, a court may order seizure and destruction or other disposition of infringing copies and equipment used in their manufacture (PTO 1991: 7).

In the United States, the Lanham Act provides a civil cause of action in federal courts based on trademark infringement, unfair competition and false advertising (Tulchin, Klapper and Montagu 1994: 17–19). The Act also provides a cause of action in the United States for certain activities taking place outside the United States. The application of the Lanham Act to conduct outside the United States was established by the US Supreme Court Act in *Steele v. Bulova Watch Co.*, 344 US 280 (1952). It allows US commercial enterprises to sue competitors who engage in a wide range of false or misleading acts. Whenever a company deceives consumers by imitating another company's trademarks or trade names it may be ordered to pay civil damages. However, there has recently been a trend to reverse the tradition of extending the Act to foreign activities which have minimal effects on US commerce.

For managers, the weakness of all of these legal remedies lies in the fact that they merely block goods at the border rather than eliminating them at the source. Moreover, Customs does not have the power to inspect all incoming shipments, looking for products that appear genuine but are not. Customs is heavily dependent on property rights owners reporting to it on incoming shipments or problems with counterfeit products. Therefore, the responsibility of enforcing IPRs falls mainly on businesses, which should keep themselves informed of their markets and the trouble spots for their products. Some companies, like Apple Computer, have been fairly successful at monitoring areas where they have reasonable evidence of counterfeiting. When they have reasonable evidence local law enforcement is then asked to seize products before they are shipped out.

Indeed, the complex nature of trade in counterfeit or pirated goods has presented a whole set of serious challenges to management, which Robinson (1988: 148–149) has summarized briefly as follows:

1 The need for an early assessment of the commercial value of a patent. The first party to apply for a patent in one Paris Convention country has only one year of grace before applying for protection in other member states.
2 The need to establish a system to ensure continuation of a patent, trademark or copyright validity.
3 The need to establish a system to monitor important markets for infringement and to prosecute infringers.
4 The need to decide whether to develop a universal trademark or a series of national or regional marks.
5 The need to determine who should own the proprietary right – the parent or subsidiary.
6 The need to keep ready access to legal experts in international trademarks, copyright, patent and trade secrets law.
7 The need to set up a subsidiary in one of the countries signatory to the Madrid Agreement through which to register the trademark, since the United States is not a signatory and most countries do not require prior use for trademark registration.
8 The need to compare the suitability of either relying on patent protection or on keeping secrecy.

GOVERNMENT RESPONSES TO COUNTERFEIT GOODS

Unilateral acts

In recent years, the highlight has shifted from individual firms to governments in the worldwide fight to improve intellectual property protection. The US government, in particular, has been very actively involved in promoting intellectual property protection internationally. So far, it has used two major vehicles in an attempt to push other nations to change their policies: section 301 of the Trade Act of 1974, and the Uruguay Round of GATT.

In 1988, the US Congress, startled at the rampant violations against US IPRs, enacted two new provisions amending section 301 of the Trade Act of 1974. Generally speaking, section 301 is a trade retaliation statute, requiring the United States to retaliate by imposing higher tariffs or restrictions to US market access in some circumstances, and authorizing retaliation in others. The two 1988 amendments have conferred more power on the US government to retaliate for unfair trade practices, and to negotiate for the improvement of intellectual property protection.

One amendment, known as "Super 301," requires the US Trade Representative (USTR) to identify unfair priority foreign trade practices having a major impact on the US economy and identify the priority countries in which those practices occur. Once the practices and countries are identified, the US Trade Representative must undertake negotiations for their removal. If the negotiations fail to lead to improvements, the United States must retaliate against the designated countries. At the end of December 1990, the controversial Super 301 provision of the Omnibus Trade and Competitiveness Act of 1988 expired, but President Clinton revived it as part of his tough stance toward unfair trade practices. Many have argued that Super 301 has violated GATT, which already provides a framework for the resolution of international trade disputes and that the revived Super 301 would further antagonize the United States' main trading partners (Svernlov 1992: 125–132).

The second amendment, known as "Special 301," specifically targets intellectual property practices. It requires the US Trade Representative to identify the countries that deny US firms effective protection of IPRs or deny market access to US businesses that depend on intellectual property law. As with a Super 301 designation, identification triggers a section 301 investigation, which leads to retaliation (Bello and Holmer 1990: 874–880). The results of the 1994 Special 301 Review announced on April 30, 1994 placed a number of countries on the priority watch list and a regular watch list. The priority list includes the European Union, Japan, Korea, Saudi Arabia, Thailand and Turkey; on the watch list are Australia, Chile, Colombia, Cyprus, Egypt, El Salvador, Greece, Guatemala, Indonesia, Italy, Pakistan, Peru, Philippines, Poland, Spain, Taiwan, United Arab Emirates and Venezuela. While USTR Kantor did not designate any priority foreign countries on April 30, he made it clear that Argentina, China and India would be designated as priority countries with the immediate initiation of investigations if they did not make "satisfactory progress" (Hill 1994b: 25).

Although many countries believe that the section 301 provision violates the spirit of GATT, the section has had some very positive impacts on world trading practices. For example, Japan, which was targeted under Super 301, has liberalized its markets in several areas. In 1992, China agreed to improve its intellectual property protection and to join the Berne Convention, just before a deadline set for retaliation under Special 301. However, to enforce IPRs unilaterally can be very expensive and difficult. According to a study made by Feinberg and Rousslang (1990: 79–89), it would probably require a large percentage increase in existing identification and enforcement costs to reduce profits lost to infringers by 1 per cent.

Bilateral acts

In the past few years, the US government reached a number of bilateral agreements aimed at improving the protection of IPRs in the targeted countries. For example, in mid-January 1992, the United States and China reached a last minute agreement on IPRs, lifting the threat of US sanctions against China for its failure to adequately protect American copyrights, patents, trademarks and trade secrets. In a Memorandum of Understanding signed on January 17, the US Trade Representative agreed to terminate its Special 301 investigation into China's intellectual property practices in exchange for a number of commitments by the Chinese government.

In the Memorandum, the Chinese government undertook to join two international copyright conventions within the next eighteen months and to take necessary steps over the next two years to enact new laws and regulations that will expand the scope of protection for American intellectual property. Nearly all of these improvements will benefit Chinese and other foreign rights holders, not just Americans. Prominent improvements include the following:

1 Extension of copyright protection for the first time to foreign owners of software, books, films, sound recordings and other subject matter.
2 Removal of the prohibition against patenting of pharmaceuticals.
3 Protection against various forms of unfair competition practices, including misappropriation of trade secrets (Simone, Jr 1992: 9).

The Memorandum of Understanding is not a solution to all problems, but rather has set up a framework within which both sides will continue to deal with the issue. The 1995 crisis between the two countries on intellectual property issues has only proven the fact that bilateral agreement does not bring an end to all the intellectual property violations. The more difficult task is to implement the agreements.

In another example, the United States and Japan signed an agreement on August 16, 1994 that made major changes in the patent policies in both countries (Riordan 1994: C1). Under the agreement, Japan pledged to process patent applications filed by American companies and inventors faster. Prior to the agreement, the United States and Japan had been at odds over Japan's lengthy process for reviewing patent applications. Getting a patent application approval usually took five to six years, which is too long for high-tech firms having short product life cycles. The agreement also calls for Japan to end the practice of allowing third parties to oppose a competitor's patent before it is granted. Japan also said that it would cancel a requirement for compulsory licensing of patents in some industries and allow foreign investors to file their initial applications in English.

For its part, the United States promised to publish pending patent applications eighteen months after they are filed. That would protect Japanese companies from so-called submarine patents, patent applications that remain secret until after a competitor goes forward with similar development plans. In the past, submarine patents could sit for decades at the Patent Office until a technology became widely used and could be used to torpedo competitors. The United States does not publish applications that are turned down, and publishes others only after they are awarded. Although most patent applications in the United States are processed within nineteen months, submarine patents usually took much longer.

Like the Memorandum of Understanding between the United States and China, the latest agreement does not mark the end of United States–Japan patent negotiations. There are many differences between the two countries; and the agreement is yet to be implemented. For example, the United States would still like to see Japan interpret patent clauses less literally. For its part, Japan wants the United States to issue patents on a first-to-file basis, as opposed to the first-to-invest criteria.

Multilateral acts

The United States has tried successfully to push GATT into a multilateral agreement on intellectual property. Among the items on the agenda for the Uruguay Round were Trade Related Intellectual Property (TRIPs) discussions (Hill 1994a: 10–11), which has been included in the recently signed GATT agreement. TRIPs requires contracting parties to GATT to provide certain minimum levels of IPRs and enforcement. It stipulates that product and process patents should be available in all fields of technology. It also specifies that patent owners be given the right to prevent others from making, using, offering for sale, selling or importing products covered by a product patent, or from using a patented process, or offering for sale, selling or importing the product obtained directly from the use of the patented process. The right to assign or license a right under a patent is also guaranteed. Disputes involving the TRIPs agreement will be handled by the General Council acting as the dispute settlement body.

However, there are some drawbacks for the United States. As a trade-off, some have argued that "Special 301" will be disarmed. While the United States has not publicly renounced the making of unilateral threats of trade sanctions because IPR protection is inadequate, such unilateral action will be illegal under GATT. Moreover, a country subject to any such threat can raise the issue before GATT and may be authorized to impose corresponding trade sanctions on the United States (Wineburg and Jarbovsky 1994: 32–33). TRIPs also prohibits the United States from discriminating between itself and a foreign country on the issue of where the invention was made when determining who has priority of invention and the right to obtain a patent. The United States will also have to extend the terms of patent protection from seventeen years to twenty years from filing of the first complete patent applications.

The United States has also successfully persuaded both Canada and Mexico to harmonize their intellectual property protection laws to that of the United States (Weiser 1993: 671–689). Part Six of the North American Free Trade Agreement (NAFTA), "Intellectual Property," represents a major victory of the United States and sets a clear example for other countries which want to have a free trade agreement with the United States.

GRAY MARKET ISSUE

The gray market issue is another challenge that intellectual property owners have to deal with to preserve their property rights. The problem is that under some circumstances legitimate goods will flow into a market through unauthorized channels. Such goods are commonly called gray market goods, i.e. foreign-manufactured goods bearing a trademark, collective mark or certification mark identical to one used or registered in the United States that are imported without the consent of the domestic mark's owner.

Gray market goods crop up either because a seller has set different prices for identical goods in different markets or because currency values fluctuate, making it profitable to purchase goods in other markets and import them. The distribution of genuine products outside manufacturers' authorized channels is big business. In 1988, for example, an estimated $7 to $10 billion in goods entered the US market through gray markets. Although they are not illegal in most cases, gray markets create complications for licensed dealers and the companies that supply them (Lansing and Gabriella 1993: 313–337).

The need for protection against the importation of gray market goods is best evidenced in cases where the gray market goods have substantially different characteristics from goods manufactured domestically carrying the same trademarks. Some gray market products do not conform to government-set standards. Others do not come with the warranties or services that are normally offered by the authorized distributor. Still others may be significantly inferior in quality. These discrepancies show that gray market goods can be potentially detrimental to a trademark owner's reputation when such goods are associated with the owner's goods. In the United States, the Lanham Act can be used to protect domestic trademark holders' rights from violation by gray market importers (Yosher 1992: 1363–1390).

The same laws that block the import of counterfeit or pirated goods may also prevent the import of some gray market goods. Many legal systems are beginning to encounter gray market problems. The United States has struggled with gray market goods for seventy years. As discussed earlier (see p. 116), by section 526 of the Tariff Act of 1930, the owner of a trademark may bar imports that bear an identical mark. The statute can also be used to allow US trademark owners to bar all gray market goods by simply denying permission for their import. However, the Customs Service over the years has interpreted the statute in a way that permits many gray market imports.

Gray market problems were eventually addressed by the Supreme Court in 1988, when a case was brought by a trade association and by Cartier, Inc., against K Mart and 47th Street Photo, Inc., two of the United States' largest gray market importers. In *K Mart v. Cartier*, 198 US 1811 (1988), the Supreme Court made a rather confusing ruling with different majorities for different parts of the decision.

Since then, this ruling has provided some general guidelines for gray marketers.

The Court laid down five fundamental frameworks for gray market imports, and for each, ruled on whether section 526 required Customs to exclude the goods unless the US trademark owner authorized the import:

Case 1 A US firm purchases the rights to register and use a foreign firm's trademark in the United States, selling the foreign firm's products in the United States. In this case, the Court held that imports of the same goods by the foreign manufacturer or by a third party who has purchased the goods from the foreign manufacturer would unfairly jeopardize the value of the US trademark holder's investment. Thus, section 526 requires the Custom Service to exclude imports in this case.

Case 2A A foreign firm manufactures goods overseas. A US subsidiary of the firm registers the foreign trademark in the United States. The Court ruled that Customs could allow the gray market goods to enter the United States.

Case 2B This case is the reverse of 2A. Here, a US firm sets up a foreign subsidiary to manufacture and sell trademarked goods. Again, the Court ruled (by a different majority) that Customs could allow the goods to enter the United States.

Case 2C In this case, the US company establishes a branch or a subdivision to manufacture goods offshore. The Court ruled that these goods were not "of foreign manufacture," as the statute required, so Customs could allow the goods to enter.

Case 3 Here, a US holder of a US trademark authorizes a foreign manufacturer to make goods and use a trademark in foreign markets. That manufacturer or a third party then imports the goods. The Court ruled that section 526 required the exclusion of those imports, unless the US trademark holder consented to the import.

The most direct impact of the several different votes on the gray market scenarios is that gray market imports are restricted only to a limited extent. If there is common control between the United States and the foreign firm, as parent, subsidiary or branch, imports may be allowed. But if the US and foreign businesses are independent of each other, the US trademark holder has the right to block unauthorized imports.

Lever Brothers Company v. United States

981 F.2d 1330 (D.C. Cir. 1993)

Facts Lever Brothers Company (Lever US), a US company, and its British affiliate, Lever Brothers Ltd (Lever UK), both manufacture deodorant soap under the "Shield" trademark. While the trademarks are registered in each country, the products have been formulated differently to suit local tastes and circumstances. Unlike the British product, the US soap lathers more, smells different, has a bacteriostat that enhances its deodorant properties, and contains colorants that have been certified by the FDA. Further, the packaging of the two soaps is somewhat different. Lever US complained that the unauthorized influx of the British soap created substantial consumer confusion and deception in the United States in violation of section 42 of the Lanham Act. However, the Customs Service permitted importation of the British soap under its common control exception, which provided that foreign goods bearing US trademarks are not forbidden when the foreign and domestic trademark owners are subject to common ownership or control. Lever US brought suit, arguing that the Customs Service's common control exception violates the terms of the Lanham Act when the foreign-made goods are materially different from their US counterparts.

Issue Should the Customs Service be required to exclude the British soap from the United States?

Decision Yes. There is nothing in the language or legislative history of section 42 suggesting that Congress intended to allow third parties to import physically different trademarked goods that are manufactured and sold abroad by a foreign affiliate of a US trademark holder. Customs argued that the British soap did not "copy or simulate" the US trademark because a trademark applied by a foreign firm subject to the common control of the domestic trademark owner is by definition genuine, regardless of whether or not the goods are identical. This argument is fatally flawed. It rests on the false premise that foreign trademarks applied to foreign

> goods are genuine in the United States. Trademarks applied to physically different foreign goods are not genuine from the viewpoint of the American consumer. Thus, section 42 of the Lanham Act bars the importation of physically different foreign goods bearing a trademark identical to a valid US trademark, regardless of the validity of the trademark's genuine character abroad or the affiliation between the producing firms.
>
> (Richards 1994: 353–354)

The case of *Lever Bros. v. United States* (1993) signifies the most recent round of litigation to address the gray market controversy. The Court in that case became the first in the gray market area to attempt seriously to link mark owner's access to the import exclusion provisions of section 526 of the Tariff Act of 1930 and section 42 of the Lanham Act. In the case, the "genuine goods," "materially differing goods" and "common control" doctrines were defined, and the appropriateness of applying unfair competition law was examined.

SUMMARY AND CONCLUSION

This chapter has discussed responses of businesses and governments to counterfeiting and piracy. In spite of difficulties in protecting IPRs, both businesses and governments can make major contributions to the improvement of IPR protection by adopting a more proactive approach. Businesses should set up an effective system of their own to monitor the flow of counterfeit goods and keep the relevant institutions of their governments well informed. They may even enlist the help of the government of the country where counterfeit goods are manufactured. On the other hand, governments can significantly strengthen the struggle against counterfeiting by improving their own systems of protection, vigorously enforcing protection laws and building up bilateral and multilateral frameworks of protection.

Although serious problems exist in protecting intellectual property rights abroad, hundreds of thousands of firms have found the transfer of such rights to be very profitable. The transfer of these rights has continually grown all over the world. With various bilateral and multilateral agreements, worldwide enforcement of protection is also significantly improving. But to eliminate intellectual property rights violations is a long-term fight, if not impossible. In this fight, businesses should take on more responsibility by improving their own management. As Robinson has commented:

> In the final analysis, the real protection of transferred technology lies in an untarnished corporate name, a high level of credibility in the eyes of consumers in all markets, superior technology supported by a record of ongoing innovation, a capability of effecting international technology transfer at relatively low cost, a design-engineering sensitivity to the requirements peculiar to different markets, and sufficient organizational flexibility to select the most effective transfer vehicle and mount appropriate controls.
>
> (Robinson 1988: 149)

Part III

International technology licensing

9 International licensing

INTRODUCTION

International licensing, defined broadly, comprises a variety of contractual arrangements whereby domestic firms (licensors) sell their intangible assets or property rights (patents, trade secrets, know-how, trademarks and company name) to foreign firms (licensees) in return for royalties and/or other forms of payment (Root 1994: 107). The transfer of these intangible assets or property rights is the core of a licensing agreement. Under this arrangement, the firms typically provide a limited right to produce and market the product in a specified geographical region. The transfer is usually supported by technical services to ensure the appropriate exploitation of the assets. Licensing agreements are normally long-term arrangements that may require significant investment by the licensee.

Licensing to foreign companies has long played an important role in business strategies in developed countries. With Japanese companies alone, American firms signed approximately 32,000 licensing agreements between 1952 and 1980. During the late 1980s, licensing fees and royalty payments brought to US licensors more than $12 billion a year, roughly twice the rate earned a decade earlier. Into the 1990s, the pace of international licensing is accelerating, with more rapid growth of international licensing led by smaller firms in industries protected by patents, such as biotechnology, semiconductors and pharmaceutical companies. International licensing is often conducted in the form of a cross-licensing agreement or technology swap between firms or as part of a larger overall strategic partnership between firms, which may involve joint ventures or other strategic alliances (James and Weidenbaum 1993: 29–30). This chapter reviews the theoretical arguments on licensing, describes key modes of technology licensing, and analyses the advantages and disadvantages

9 International licensing

INTRODUCTION

International licensing, defined broadly, comprises a variety of contractual arrangements whereby domestic firms (licensors) sell their intangible assets or property rights (patents, trade secrets, know-how, trademarks and company name) to foreign firms (licensees) in return for royalties and/or other forms of payment (Root 1994: 107). The transfer of these intangible assets or property rights is the core of a licensing agreement. Under this arrangement, the firms typically provide a limited right to produce and market the product in a specified geographical region. The transfer is usually supported by technical services to ensure the appropriate exploitation of the assets. Licensing agreements are normally long-term arrangements that may require significant investment by the licensee.

Licensing to foreign companies has long played an important role in business strategies in developed countries. With Japanese companies alone, American firms signed approximately 32,000 licensing agreements between 1952 and 1980. During the late 1980s, licensing fees and royalty payments brought to US licensors more than $12 billion a year, roughly twice the rate earned a decade earlier. Into the 1990s, the pace of international licensing is accelerating, with more recent growth of international licensing led by smaller firms in industries protected by patents, such as biotechnology, semiconductor and pharmaceutical companies. International licensing is often conducted in the form of a cross-licensing agreement or technology swap between firms or as part of a larger, overall strategic partnership between firms, which may evolve into joint ventures or other strategic alliances (James and Weidenbaum 1993: 29–30). This chapter reviews theoretical arguments on licensing, describes key modes of technology licensing and analyzes the advantages and disadvantages of licensing for both licensor and licensee. At the end of the chapter the changing environment of the former Soviet Union and Eastern Europe and its impact on licensing development in the region is discussed.

THEORETICAL REVIEW OF LICENSING INTERNATIONALIZATION

As has been discussed in the introduction, the owner of specific technology has certain monopolistic power. Exchanges of technology do not follow perfect market rules and serious imperfections exist. Owing to such imperfections, owners may take either of two alternative courses of action. They may refuse to transfer technology, because they do not see adequate returns and would rather fully utilize the technology themselves. This is the so-called internalization of technology (Buckley and Casson 1976). Alternatively, they may decide to license the technology, though imposing some restrictions on how the technology transferred is to be used, such as geographical restrictions (Teece 1981: 81–96). This is the so-called externalization of technology.

What are the options available to the owner for exploiting the technology in the global market? Should the owner focus on internalization or should it externalize? To answer these questions, Rugman *et al.* (1985: 126–130) use the following variables to study the preference among multinational corporations (MNCs) for international market exploitation modes:

Country-specific costs

C = aggregate production function (home)
C^* = aggregate production function (foreign)

Special costs

M^* = Export marketing costs (goods market)
A^* = Additional cost of FDI (goods and factor markets)
D^* = Risk of dissipation of FSA (firm specific advantage) (risk of lost sales and costs of enforcement of licensing agreement)

The MNC is faced with the following alternatives for servicing foreign markets:

1 Export if $C + M^* < C^* + A^*$ (exporting is cheaper than FDI) and $C + M^* < C^* + D^*$ (exporting is cheaper than licensing).
2 FDI if $C^* + A^* < C + M^*$ (FDI is cheaper than exporting) and $C^* + A^* < C^* + D^*$ (FDI is cheaper than licensing).
3 License if $C^* + D^* < C^* + A^*$ (licensing costs are less than FDI) and $C^* + D^* < C + M^*$ (licensing costs are less than exporting).

For Rugman *et al.*, if markets are perfect (i.e. with no information cost or barriers to trade), exporting should be the first option. If markets are imperfect (i.e., with tariffs or other trade barriers) and if the risk of dissipation of FSA is high, foreign markets should be served via FDI. However, if the host government imposes regulations on the MNC that are more costly than the benefits of FDI and the MNC does not face the risk of dissipation, then it should license its technology (FSA) to a foreign licensee. Because of the complexities in international operations and a variety of external as well as internal constraints on MNCs, a combination of various market entry strategies such as licensing, exports and FDI may be adopted.

Licensing may also be applied to take advantage of the ownership-specific advantage (technology), by allowing the licensee to use the technology under certain conditions for a specific period of time, for a specified price (license fees or royalties). Another option that is frequently used by the owner of technology is to form a joint venture, which is an intermediate option between externalization and internalization. While the owner of technology may wish to exclusively exploit the technology (internalization or FDI), various barriers and regulations in different nations may compel the owner to consider a joint venture strategy.

In order to understand the desirability of using a licensing strategy versus an FDI mode to transfer technology, Contractor (1985: 277–320) has done a theoretical exploration to compare their costs and compensation (Figure 9.1). Assume that two categories of costs C_1 and C_2 are incurred completely in the technology supplier's country. C_1 signifies the present value of sunk R&D costs of this particular technology, and assume that C_1 is the same regardless of the strategy used to commercialize it overseas. C_2 signifies the present value of general administrative headquarters overheads for all technologies and products (including the cost of failed R&D efforts). T_1 is the present value of the costs of transferring technology to the other nation, and T_2 is the present value of repatriated earnings or returns on the technology from the nation. Whereas C_1 represents the cost of

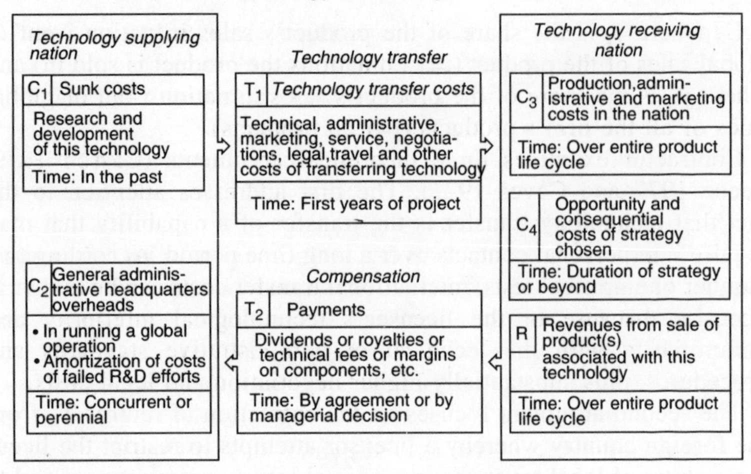

Figure 9.1 Schematic of international technology transfer costs and compensation in either licensing or direct investment
Source: Contractor 1985: 281 (reprinted by permission of Greenwood Publishing Group, Inc., Westport, CT)

producing the technology, T_1 refers to the cost of reproducing it in the foreign country, i.e., the variable cost of technology.

After the technology is transferred, a product will be manufactured and marketed at cost C_3, to earn revenues R in the recipient nation. The present value of $R - C_3$ over the product life cycle in that nation represents the gross margin on the technology from that nation. C_4 symbolizes the opportunity costs of either strategy, which refer to the lost profits (or savings) that would have accrued without using the strategy, such as the profits on future direct sales to that territory that will now be eliminated. With licensing, C_4 also indicates consequential costs, which refer to loss of profits that may result from the fact that on expiry of the license, or even before, a licensee may compete with the licensor in third countries.

Axiomatically, licensing fees or dividends repatriated have to be derived from the gross margin on the technology, or $T_2 <$ or $= R - C_3$. But to decide between licensing and FDI, in Contractor's view, the focus has to be laid on the net technology margin from the country, or $T_2 - T_1 - C_4$. To focus only on the net direct cash flow ($T_2 - T_1$) may result in wrong decisions. This is particularly so in companies with an international division or licensing section as a separate profit center to which $T_2 - T_1$ accrues, while the opportunity costs C_4 are borne by other product divisions. To cover all relevant costs, the company would like to have $T_2 >$ or $= (T_1 + C_4) + y_iC_1 +$

$y_{ij}C_2$ where y_i is the share of the product's sales in nation i out of global sales of the product (i=1...n nations the product is sold in) and where y_{ij} is the share of the product's sales in nation i out of global sales of all the firm's products (j=1...p products).

Contractor examines three theoretical arguments (Casson 1979, Teece 1977 and Caves 1971). The first addresses attention to the fact that technology transfer is the transfer of a capability that may require interpersonal contacts over a long time period. According to a number of empirical tests, international transfer costs can be higher in licensing because of the licensee's technological inferiority and unfamiliarity with the technical or administrative standards and procedures, plus substantially higher negotiation and legal costs.

The second argument focuses on the extraction of returns T_2 from the foreign country whereby a licensor attempts to restrict the licensee's *share* of local revenues to a normal return on investment and to maximize its own profit. But the licensee tends to place a lower value on the technology so as to provide license fees below the repatriable profit that may be earned by an equity investment ($T_2 < R - C_3$). Moreover, when licensing the same technology in several countries, the licensor may be forced to use a uniform price for all licensees, a suboptimal option compared to the discriminatory compensation possible in global equity investments. Unlike licensing payments, dividends are not subject to time limit.

The third argument is based on the monopoly power of the MNCs in product markets overseas, with an emphasis on the product revenues R. For the same production and distribution costs C_3, an MNC usually obtains higher prices (R) compared to a local licensee for a number of reasons, including superior product quality, superior organization, international brand recognition, xenophilia, etc. Therefore, the margin ($R - C_3$) earned from the product market is very likely to be higher in FDI than that earned by a licensee.

Contractor concludes that the net technology margin extracted from a country ($T_2 - T_1 - C_4$) should be more often than not lower in licensing than from a company's own subsidiary because the returns T_2 from licensing are likely to be lower while transfer costs T_1 and opportunity costs C_4 are usually higher. Licensing may provide higher or equivalent margins under conditions where the technology is standardized, simple or mature, where the licensee has compatible technology competence or is on the same technological level as the licensor, or where the opportunity costs of licensing are low or zero, as in some developing countries which restrict both import and FDI.

MAJOR MODES OF TECHNOLOGY LICENSING

There are basically four modes of technology licensing (Shahrokhi 1987: 35–36): inventor–corporation licensing, corporation–corporation licensing, cross-licensing and intrafirm licensing.

Inventor–corporation licensing

The inventor/patent holder is an individual, or entrepreneur or a small firm. Owing to the shortage of financial resources, the inventor is not in a position to exploit the invention commercially. Subsequently, the invention needs to be licensed to a company capable of applying it. Under such circumstances, the licensee may gain full control over manufacturing and marketing of the product in exchange for the obligation to pay the licensor the agreed licensing fees. The arrangement can work well if a proper agreement that protects the interests of the licensor is formed. The licensor should carefully define the obligations of the licensee, its agents and its employees to protect the confidential nature and acknowledge ownership of the intellectual property being disclosed pertaining to the license agreement (Sherman 1992: 42).

Corporation–corporation licensing

This is probably the most typical licensing situation in which a corporation licenses its technology (patent rights and/or technical know-how) to another corporation in exchange for a technology fee and a royalty (a percentage of the sales). In this situation, licensing is one of many options available to the licensor (others include joint venture and FDI). A company may choose to license out of a number of motivations discussed in this chapter.

Some companies have consistently used licensing as the preferred form of technology transfer. The main motivations may vary, such as a quick recovery of expensive R&D costs, or quick market penetration without making a heavy capital investment overseas, etc. However, one common ground for these companies to regularly license their technologies is that the R&D program in such companies were so successful that they have confidence in retaining a long lead in product development over their licensees, thereby minimizing the risk that the technology transferred will produce a fierce competitor or that the licensees may want to terminate the relationship at the end of the licensing period. The aggressive semiconductor technology licensing program (related to basic technology patents) of Texas Instruments Inc. generated most of the $2 billion in royalties collected by the company from 1987 to 1994 (Lineback 1994: 26).

Cross-licensing

This arrangement frequently occurs between two companies of similar standing in terms of technological capability. The motivation is to be able to share the new development of the licensed technology rather than to reap more royalty income. In such a situation, the licensor will not ask the licensee to pay for the technology, but instead will seek the exchange of the licensee's R&D findings for new developments. However, some cross-licensing agreements might arise from a licensee's insistence that the original licensor accept a license instead of payment. In this case, the motivation for licensing may not be the original licensor's interests in access to technology. Cross-licensing might also result from infringement suits, where a mutual exchange of licenses becomes part of the settlement (Telesio 1979: 64).

Cross-licensing is a popular arrangement in many industries (for example the pharmaceutical industry) where technology is very advanced and R&D costs can be extremely high. Not uncommonly, cross-licensing may involve a whole series of exchanges of technologies between the companies in similar industries. Toshiba Corp. and Seagate Technology Inc., for example, signed a broad, worldwide patent cross-licensing agreement which permits each partner to use a number of magnetic mass storage technologies covered by the other's patents (*Japan*, July 21, 1994: 89).

Intrafirm licensing

This is a type of vertical licensing arrangement whereby technology is transferred between the parent and its overseas subsidiaries. Intrafirm licensing is a fairly popular licensing mode among MNCs. One motivation may be related to the attempt to minimize political risk in the host country of the MNC's subsidiary. A nationalistic government, for example, may impose control over transfers of certain types of funds. Other motivations may be legal and financial, such as the transfer of pricing to avoid taxes. In some cases, MNCs allocate the parent's R&D expenditures to their foreign subsidiaries, in order to boost the morale of or take advantage of R&D competence in their subsidiaries.

MAJOR ADVANTAGES AND DISADVANTAGES OF TECHNOLOGY LICENSING

Advantages

There are a number of advantages that may motivate the licensor to license out its technology. The primary advantage of licensing out is to enable firms to earn additional profits on existing products or technologies without making substantial new investments, either in marketing or production (Telesio 1979: 13). This advantage is especially beneficial to smaller firms that do not have sufficient financial and managerial resources to establish a more ambitious overseas presence. An uncertain sales potential in the target market may increase the attractiveness of licensing out as compared to direct export and equity investment. If the licensed product sells well, the company may decide to enter the market by either directly exporting or setting up production facilities in the target country.

Moreover, licensing-out arrangements can be made with foreign companies in return for market entry when barriers exist. These barriers may include import restrictions and tariff barriers, local political or industrial pressures, local distribution systems discriminating against foreign-made products, and heavy transportation costs. Instead of exporting a physical product, the manufacturer sells intangible assets and services that are not subject to import restrictions. Usually, manufacturers license out when exports are not possible to a target market because of the imposition of tariffs or quotas or when exports are no longer profitable as a result of more intense competition (Branch 1990: 96). In some industries, such as aerospace, licensing may open up the market for components from US manufacturers that are needed to manufacture the licensed products.

Another advantage of licensing is the relatively low political risk compared with an equity investment. Many host governments prefer licensing over foreign investment as a channel to obtain technology. In addition, licensing is not subject to expropriation, since the licensor does not own physical assets in the target country. If the worst occurs, the most that a manufacturer can lose is the licensing income. Although this may not be insignificant, the loss is usually well below what would be incurred in the case of a manufacturing subsidiary.

There are also a number of advantages that may motivate the licensee to adopt license-in strategy. However, very few scholars have researched the licensees' motivations for acquisition of technology through licensing arrangements. The decision to obtain technology by a firm has always come from market considerations. No one would be likely to buy a technology without intending to use it either in the laboratory or in the market (whether to sell more competitive products or to monopolize the market by controlling the patent). Moreover, many licensees are not necessarily the simple followers of their licensors in the relevant technological fields.

Licensees may have any of the following motivations to acquire technology via license (Killing 1980: 38–46):

1 To obtain the needed technology to improve the quality of existing products and/or to supplement their own in-house R&D activities.
2 To launch a new product without having to take the risk of spending heavily on R&D which may end in failure.
3 To obtain the right to operate or to settle a patent dispute (the licensee might have developed the technology independently of the licensor, but because the licensor has already secured a patent on the technology, it has to acquire a license).
4 To get the technology (product) to the market faster (although the licensee has the capability to develop the needed technology (product), obtaining a license will expedite the manufacturing and marketing process).
5 To penetrate new geographic markets which are more open to goods produced with licensed technology.
6 To take advantage of the market protection that the local governments provide in favor of local manufacturers.

Disadvantages

Prior to licensing its technology, the licensor should attempt to foresee all the potential problems that might arise through the licensing agreement. A major disadvantage of licensing is loss of control (Telesio 1979: 15). Once a contract has been signed, the foreign licensee controls the manufacturing, marketing and distribution of the product. Even when the licensor makes a good selection of its licensee, it remains dependent on the licensee's performance. It can do very little to compel a better performance short of terminating the license. This worry has deterred many companies from forming international cooperative contractual relationships.

When a licensee is not committed to a good compliance with the contract, the field of use constitutes one major source of conflict between the licensor and the licensee. The licensee may use technology in fields not specified or paid for in the contract. Geographical restriction (if there is any in the contract) is also frequently violated, due to the highly regulative nature of antitrust laws of many countries and the frequent interference of host governments on behalf of local licensees (see Chapter 12). The licensee may also exploit markets not included in the contract. Although the licensing agreement may have specific provisions for such problems, it is very costly and difficult to maintain full control or to police the licensed technology.

Moreover, the brand or company names may suffer severe damage as a result of poor production quality by the licensee. Although some control problems can be ameliorated via strict quality guidelines put into the contractual agreement, these agreements may entail long-winded and complex negotiations to ensure that the licensing firms are properly compensated. Again, even after the agreement is signed, compliance is difficult to monitor and enforce. This is particularly true in countries where the legal system does not provide adequate contractual protection. One option to improve enforcement is the threat not to renew the licensing arrangement for licensed technologies that have good prospects of improvement and product line extensions (Davidson 1982: 55–56).

The creation of a future competitor for the licensor is another major disadvantage. The licensee will be competing with the licensor's product not only in the geographical markets covered in the contract but in other markets as well. The problem can be very severe after the expiration of the license. This is an opportunity cost to the licensor, and should be given due consideration in royalty and compensation arrangements. While there may be provisions for such opportunity costs, the licensor cannot be fully compensated for the existence of new competition. Especially if the licensee achieves technical and marketing competence combined with its other advantages (e.g., location-specific advantages), it becomes very difficult for the licensor to cope with challenges from the licensee.

Some may argue that strong patent protection may substantially reduce the danger of such competition. However, many companies that have entered into licensing arrangements with competitors or potential competitors have bitterly complained about this competition. Examples in which former licensees become fierce competitors or even squeeze their licensors out of the market are numerous. For instance, several of the Japanese color television manufacturers that have successfully entered the US market are former licensees of RCA. Yamaha's penetration of the US musical instruments market took place after the expiration of patents held by leading US firms (James and Weidenbaum 1993: 32).

Another major disadvantage is the absolute size of income derived from a licensing agreement. Licensing generally produces lower revenues than manufacturing locally, even when licensors can charge an optimal licensing fee. As a common practice, royalty rates seldom exceed 5 per cent of sales, and are usually limited by rates in a licensor's prior licensing agreements. Moreover, unlike export or investment revenue, licensing agreements are usually limited to five to ten years, which, if combined with a low royalty rate, may not be sufficient to recover high product development costs (Root 1994: 110).

Licensing arrangements may not be a favored option if the company already has excess domestic capacity. Unless overseas market barriers are too costly or difficult to overcome, the company would prefer to open up more export markets. If technologies are firm-specific or exhibit a large minimum efficient scale relative to total world output, the company may prefer foreign direct investment as a suitable option (Mowery 1988: 8).

Licensees also have a number of reasons not to take a license (Shahrokhi 1987: 39). These include: high opportunity costs resulting from a lengthy negotiating process; excessive adaptation costs to utilize the licensed technology; lost advantages of building up a firm's own R&D and the "not invented here" (NIH) syndrome displayed by a licensee's own R&D team; and the status quo of existing patent laws (either weak or strong laws), as well as the possible ability of the recipient's company to develop around the patented technology.

LICENSING IN THE FORMER SOVIET UNION AND EASTERN EUROPE

Tremendous changes have taken place in both the former Soviet Union and Eastern Europe since the collapse of Communism in these countries. There are many new opportunities but also tremendous challenges resulting from the uncertainty and complexities of the transition. Many laws have undergone fundamental changes, due to the radical change of the political and economic systems. In Russia, for example, the old Soviet laws have either lost their force or become irremediably outdated while the new Russian legal framework regulating intellectual property relations is still in the process of development (Ehrbar 1993: 213–219).

Like other countries in the former Eastern bloc, the existing laws of Russia have granted all enterprises the right to conclude licensing agreements directly with foreign companies or through Litsensingtor (a state-owned commercial firm). The only state institution still performing a sort of supervisory role in the field of technology transfer is Rospatent. Nevertheless, Rospatent's role is more or less restricted to state registration of technological ownership without combining policy implementation. All restrictions on the import of foreign technology have already been lifted. Foreign companies have even been awarded the right to sell a single license to various Russian enterprises, establishing a separate licensing agreement with each.

So far, no governmental institution has been given clear-cut responsibility for supervising the commercial and financial conditions of international technology trade. There is currently no legal approval process for such agreements, though approval for technology transfer was definitely required under all-Union foreign trade regulations in effect until January 1, 1992. It is generally believed that the approval requirement is likely to be reintroduced under Russian legal regulations currently still being debated in the relevant parliamentary subcommittee. If such an approval requirement is reintroduced, the approval would probably be issued by the Department of Licensing of the Russian Ministry of Science, Higher Schooling and Technology.

Since 1991, the Russian parliament has also adopted a package of laws on intellectual property protection in the first reading, which means that the laws may still be revised and will not come into effect without a presidential signature. These draft laws have for the first time classified patents as a tradable commodity, and the patentee may enjoy ownership for up to twenty years. Under the draft laws, foreign legal entities may also secure the same ownership rights as local organizations, though they may be required to register their patents or trademarks locally. There is currently a Russian profit tax law, which imposes a 15 per cent tax on royalty transfers.

Russian enterprises in the military sector have also obtained the right to exchange technologies with foreign partners. This change has allowed manufacturing enterprises to export not only MiG-29 aeroplanes and finished T-80 tanks to many countries and but also to sell the production technologies for their manufacture. Technological cooperation has also been encouraged by the government in the field of rocket science and space technology, which were under strict security control in the past.

Significant changes have also taken place in the licensing business in Eastern Europe. In the 1970s and 1980s, when direct investment in East European countries was very difficult (if not impossible), many Western companies considered licensing agreements as an important market-entry strategy and a channel for future business. Licensing offered these Western companies the advantage of small up-front investment, limited drain on management resources, and quick returns. The majority of licensing agreements were distributed in export-oriented light and consumer goods industries. Western companies sold licenses to East European partners for production items as diverse as textiles and garments, footwear, beer, cigarettes, telephone systems, electrical appliances, foam rubber and shock absorbers.

Since the opening of the region to Western trade and investment after 1989, many companies have rushed in to secure local production capacity and market share by establishing joint ventures and even acquiring East European enterprises. However, as the risks of foreign direct investment in Eastern Europe have become very obvious, Western companies have begun to rediscover licensing as a desirable alternative strategy to other forms of more risky business. Many companies are employing licensing agreements to build up their brand name in preparation for future investment. In general, licensing is flexible and can be easily integrated into other business strategies, ranging from exports to joint ventures and to wholly owned subsidiaries. The licensing business is, therefore, particularly attractive to small and medium-sized companies.

Companies which have established licensing agreements in Eastern Europe have found the following advantages: immediate income in the form of royalties and advance fees, usually based on volume; limited capital investment which is only required at the initial stage, when the licensor transfers his management and technical resources as well as provisions for quality control and maintenance to the licensee; the opportunities to increase familiarity with and develop a proven partner for deeper business involvement, including an eventual acquisition of the licensee's enterprise; and the possibility of developing business with the licensee in other product lines. Among the top concerns of Western companies is the danger of substandard products and resulting brand erosion if the licensee suffers from management and manufacturing problems (Langenecker 1993: 1).

SUMMARY AND CONCLUSION

This chapter has reviewed two theoretical studies on licensing conducted respectively by Rugman and Contractor. The former examined three major alternatives for servicing the foreign market, including export, licensing and foreign direct investment. The latter compared costs and compensation of licensing and foreign direct investment. This was followed by an introduction of four key modes of technology licensing: i.e., inventor–corporation, corporation–corporation, cross-licensing and intrafirm licensing. In the following section, advantages and disadvantages of licensing for both licensor and licensee have been analyzed in detail. Finally, there was a brief discussion of the changing legal environment for licensing in the former Soviet Union and Eastern Europe, the changing licensing system in Russia and current licensing practices of Western businesses in Eastern Europe.

In short, the success of a licensing strategy depends on a variety of factors, ranging from a careful selection of strategies to a good understanding of relevant governmental regulations. One important issue worth special attention is the need for adequate protection for the licensor's intellectual property rights. To ensure proper protection, the licensor should pay close attention to the initial selection of the licensee, which includes face-to-face talks, inspections of the recipient's facilities, financial standing and market position, and discussions with third parties. The licensor should negotiate a licensing agreement which gives the licensee sufficient incentives to protect the relevant intellectual property and should maintain constant awareness against possible breaches and abuse of confidentiality. Other important issues relate to price negotiations and the payment schedule, which is a highly complicated and controversial process, due to the nature of a technology trade. These items will be discussed in the following chapter.

10 Price negotiations in international technology licensing

INTRODUCTION

In the transactions of international technology licensing between independent licensors and licensees, pricing is usually the focus of the disputes. Many technologies are not transferred gratuitously, because the majority of them are proprietary by nature and protected under patent systems. Moreover, advanced technologies are disproportionately concentrated in some large companies in a few industrialized nations. International technology markets are generally characterized as imperfect (Buckley and Casson 1976) and transfer costs can be significant as a result of such market imperfections (Davies 1977). According to Contractor's study (1980), there are, on average, approximately five alternative suppliers of a similar technology in the world. The monopoly on the supplier side is also matched by the imperfect nature of technology recipients. Appropriate clients in recipient host countries are relatively few and tariff and/or non-tariff barriers contribute to the imperfection of the market (Contractor and Sagafi-nejad 1981).

Supplier–recipient relations in international technology markets are characterized as a "bilateral monopoly" where price is heavily influenced by convoluted bilateral negotiations. The result tends to be determined by the relative positions of suppliers and recipients. Moreover, international technology transfer usually takes place over a long period of time, because technology transfer is to cultivate a technology capacity on the part of the licensee to absorb the technology of the licensor over time. Very rarely is technology transferred (sold) to the licensee as a once-and-for-all patents and know-how exchange. Thus, the price of a technology transfer usually comprises a series of payments by the licensee to the licensor over time. This makes the pricing negotiations even more complicated. As intrafirm technology licensing follows different price structures due to the transfer pricing practices of MNCs, this chapter will focus on the general pricing principles of technology transfer between independent licensors and licensees.

COSTS AND REVENUES IN INTERNATIONAL TECHNOLOGY LICENSING

International technology licensing is possible when both the licensor and licensee of a particular technology perceive a reasonable chance of receiving net positive economic benefits from the transaction. The licensee becomes interested when he believes that the possession of a particular technology will strengthen his competitive position in the market, thereby bringing in higher profits. The licensor considers transferring the technology when the perceived compensation from the licensee surpasses costs associated with the transfer. The source of licensing revenue must be derived from the exploitation of the proprietary technology by the licensee. To determine licensing revenue, one must estimate the incremental benefits added by the transferred technology to the licensee, or incremental value.

Incremental value is embodied in one or all of the following (Yin 1993: 40–43):

1 Lowering production costs.
2 Raising quality and performance as well as sales value.
3 Increasing sales.

Incremental value derived from lowered costs When a new technology improves the productivity of the buyer's production line (i.e. reduces material consumption, thereby reducing the production costs of the buyer) but has no direct impact on the quality of the products, the incremental value generated by the transferred technology will be (time value is not yet included in the formula):

$$R = (C_0 - C_1) QN$$

Here R = the incremental value produced by the new technology
C_0 = unit cost before adopting the new technology
C_1 = unit cost after adopting the new technology
Q = annual output
N = effective years of the contract

Incremental value derived from higher product quality and performance Under certain circumstances, a technology is acquired not to lower the cost but to improve the quality and performance of the product, so that the price of the product will be higher. Assuming that production costs remain unchanged, the incremental value resulting from the transferred technology should be (time value is not yet included in the formula):

$$R = (P_1 - P_0)\,QN$$

P_0 = the unit price before the improvement
P_1 = the unit price after the improvement

If production costs also increase in the process of improving product quality and performance, the relevant incremental value will be:

$$R = [(P_1 - P_0) - (C_1 - C_0)]\,QN$$

The incremental value brought in by increased sales The brand or reputation transferred in the package of technology transfer may also help boost the sales. The formula (time value is not yet included) for the additional profits brought in by the increased sales should be:

$$R = (P_0 - C_0)\,(Q_1 - Q_0)$$

P_0 and C_0 refer to the price and costs of the product
Q_1 = annual sales after adopting the new technology
Q_0 = annual sales before adopting the new technology

If all the above factors are considered at the same time, plus the time value of the money, the formula for the additional profits should be:

$$R = \sum_{t=1}^{N} \frac{1}{(1+r)^t}\,[(P_t - C_t)\,Q_t - (P_0 - C_0)Q_0]$$

The incremental value that the licensee creates via the acquired technology is the maximum the licensor can expect to obtain. However, it is highly unlikely that any licensee would like to surrender all the incremental value to the licensor. Thus, the price a licensee accepts normally constitutes only a part of the incremental value. Licensing revenue (which includes incremental value plus the licensee's production and marketing costs) should be allocated to the

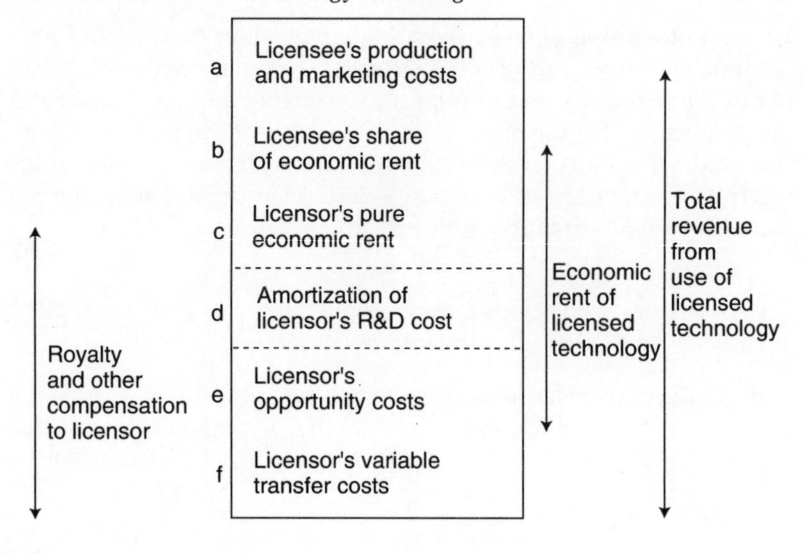

Figure 10.1 Allocation of licensee's revenues from the sale of the product using licensed technology
Source: Root and Contractor 1981: 24

licensor and the licensee to cover their respective costs and to add to their respective economic rents from the deal.

The economic rent of the licensing agreement refers to the licensee's total revenue from the use of the licensed technology minus the sum of the licensee's production and marketing costs and the licensor's transfer costs: (a + b + c + d + e + f) − (a + f) = (b + c + d + e). The licensor's compensation has to be allocated to three types of costs before the pure rent accruing to the licensor from the technology can be determined. The three types of costs are: the licensor's transfer costs (incurred in transferring technology to the licensee and all the ongoing costs of maintaining the agreement), the R&D cost of the licensed technology and opportunity costs. Figure 10.1 designates R&D costs in the licensor's share of the economic rent, because licensing revenues may be possibly used to amortize R&D costs of the licensor.

Without counting O_1 and O_2, the licensor's contribution margin from the agreement can be understood in the following formula (Contractor 1981: 34):

Table 10.1 Categories of returns and costs of technology supplier

Returns to supplier firm in year t	*Transfer costs incurred by supplier firm in effecting the agreement in year t*	*Other categories of costs*
RET_{1t}: Front-end or lump sum fees	$COST_{1t}$: Technical services (direct and overhead)	O_1: Total of sunk or developmental costs for the product or process transferred, up to inception of agreement
RET_{2t}: Royalties	$COST_{2t}$: Legal costs (direct and overhead)	O_2: Opportunity costs (for example, losing export sales or direct investment opportunities in licensee's country or in other territories)
RET_{3t}: Technical assistance fees	$COST_{3t}$: Marketing assistance to recipient	
RET_{4t}: Fees for other specific services rendered	$COST_{4t}$: Travel and management personnel costs (not included above)	
RET_{5t}: Payment in equity of recipient firm and dividends thereon	$COST_{5t}$: Other direct costs associated with executing agreement	
RET_{6t}: Net margins and commissions on materials or goods supplied or received		
RET_{7t}: Value of grantbacks (improvements or innovations made by licensee)		
RET_{8t}: Tax savings arising from arrangement, if any		

Source: Contractor (1981: 35)

$$\text{Present value of contribution margin} = \sum_{t=0}^{n} \frac{\sum_{i=1}^{8} RET_{it} - \sum_{j=1}^{5} COST_{jt}}{(1 + r)^t}$$

While the relative importance of these three types of costs may vary in different licensing agreements, transfer cost was identified in the study by Root and Contractor (1981) as more important among US firms in arm's-length licensing agreements with foreign licenses. One reason might be based on the fact that it is very difficult for a multiproduct MNC to allocate joint R&D costs incurred over years to a particular agreement. Therefore, US firms usually amortize R&D costs only over domestic and foreign affiliate sales. Opportunity costs of the licensing agreement are viewed as less important among US firms, because they are highly hypothetical and are not properly understood. For these reasons, the licensor is usually flexible about these costs in technology licensing negotiations. For similar reasons, the licensee is reluctant to recognize these costs as legitimate, at least at the level of the licensor's estimation.

TECHNOLOGY PRICING PRINCIPLES

Traditionally, business marketing managers have based their pricing strategy on three principles: costs, probability and demand. Costs refer to a minimum level for setting price in order to break even, thereby representing a beginning point in pricing. Costs set the floor below which the company markets its products at a loss. Costs indicate company effectiveness and efficiency but do not reflect the amount customers are willing to pay. Therefore, the use of cost-based pricing may lead to a price level deviating from what the market is willing to pay (Nagle 1987). The probability principle is based on the assumption that the lowest bid submitted will be accepted. One major problem with this principle is the business marketing manager's limited ability to determine the appropriate range of prices and to estimate the best probabilities (Bingham and Faffield 1990). According to the demand principle, the customer's price sensitivity to demand sets a ceiling to the price beyond which the customer will not pay. Managers applying this principle have low interest in niche strategies and differentiation and set a price equivalent to that of the competitor's (Kotler 1991).

In spite of those flaws, the three traditional pricing principles remain applicable in present business marketing operations. In technology transfer, the ideas contained in the three principles are also useful in understanding the dynamics of pricing. However, technology transfer does have its own unique features. One very important principle for pricing in technology transfer is that both the licensor and licensee share the incremental value generated by the technology. The United Nations Industiral Development Organization (Arni 1984; UNIDO 1983) interprets this as the principle of LSLP (or Licensor's Share of Licensee's Profit) and creates a formula to compute LSLP:

$$\text{LSLP (\%)} = \frac{\text{The fee received by the supplier (technology price)}}{\text{The profit of the recipient}} \times 100\%$$

Technology price = LSLP (%) x the recipient profit

There are two ways to decide LSLP. The first is LSLP based on the rate of return. The rate of return for the licensee is used to show the profitability of the acquired technology and indicate a proper proportion of the technology fee in the recipient's profit. According to studies of UNIDO, the rate of return should be between three and five, i.e. each $1 technology fee should have at least a profit of $3 to $5 in order to qualify as a worthwhile project. The proportion between the profit and technology fee should be 3 to 5: 1. In other words, the technology price should be 20 per cent to 33 per cent of the profit (LSLP).

$$\text{Rate of return} = \frac{\text{total profit}}{\text{total technology fee}}$$

The second approach is LSLP based on commonly adopted figures. UNIDO once collected statistics and did analysis on the prices of technology acquisition contracts made by India and other countries. By placing LSLP between 16 per cent and 26 per cent, it suggested that LSLP should not exceed 30 per cent. In the conference of the Licensing Executive Society held in 1972, most representatives considered 25 per cent as rational; and many Japanese books adopted such a figure. Most cases in the American courts put LSLP between 10 per cent and 30 per cent.

UNIDO's approach, however, has a number of drawbacks (Bidault 1989: 67–69): first, it is hardly possible to determine the profit which has actually been made, even at a later stage. The licensed technologies are usually integrated into complex activities and it is impossible to isolate them for accounting purposes. Second, as profitability changes over time, the LSLP should change correspondingly, but UNIDO's studies only indicate the significant variations in the LSLP in its samples without justifying them. Third, UNIDO's approach fails to take into account the role of the licensors' strategies in the fixing of prices (and in the determination of the LSLP). Some companies may try to maximize their LSLP while others are willing to give up part of their remuneration for the licensed technologies to facilitate commercial transactions.

The pricing issue in technology transfer is further complicated by the gaps in the pricing principles used by the licensor and the licensee. Pricing by the seller usually consists of three elements:

1 To recover the direct cost. Normally, a seller would like to recover at least the direct cost of the transfer. Otherwise the transfer becomes a donation.
2 To recover part of sunk cost and opportunity cost. The rate of recovering these two costs varies a great deal, depending upon the specific situation of this technology and market.
3 To deduct a percentage from the incremental value.

Therefore, the total price should be:

$$P = b_1C_{direct} + b_2C_{sunk} + b_3C_{opportunity} + b_4R$$

(b_1 to b_4 are the relevant coefficients, ranging between 0 and 1)

In technology transfer, the licensor is usually the bidder and quotes price first. The licensee usually has the advantage of providing a counter-offer, although he must know the real value of the technology he is prepared to buy in order to determine his own ceiling price and floor price. In determining whether to acquire a technology, three conditions should be considered:

1 The total cost of acquiring technology should be lower than the buyer's R&D cost on the technology.
2 The incremental value brought in by the acquired technology should be larger than zero.
3 Where similar technology is offered by several sellers, the buyer should choose the one making the best offer.

To maximize his share of economic rents, the licensor tends to estimate a higher incremental value. Normally he follows three principles in determining the ceiling price:

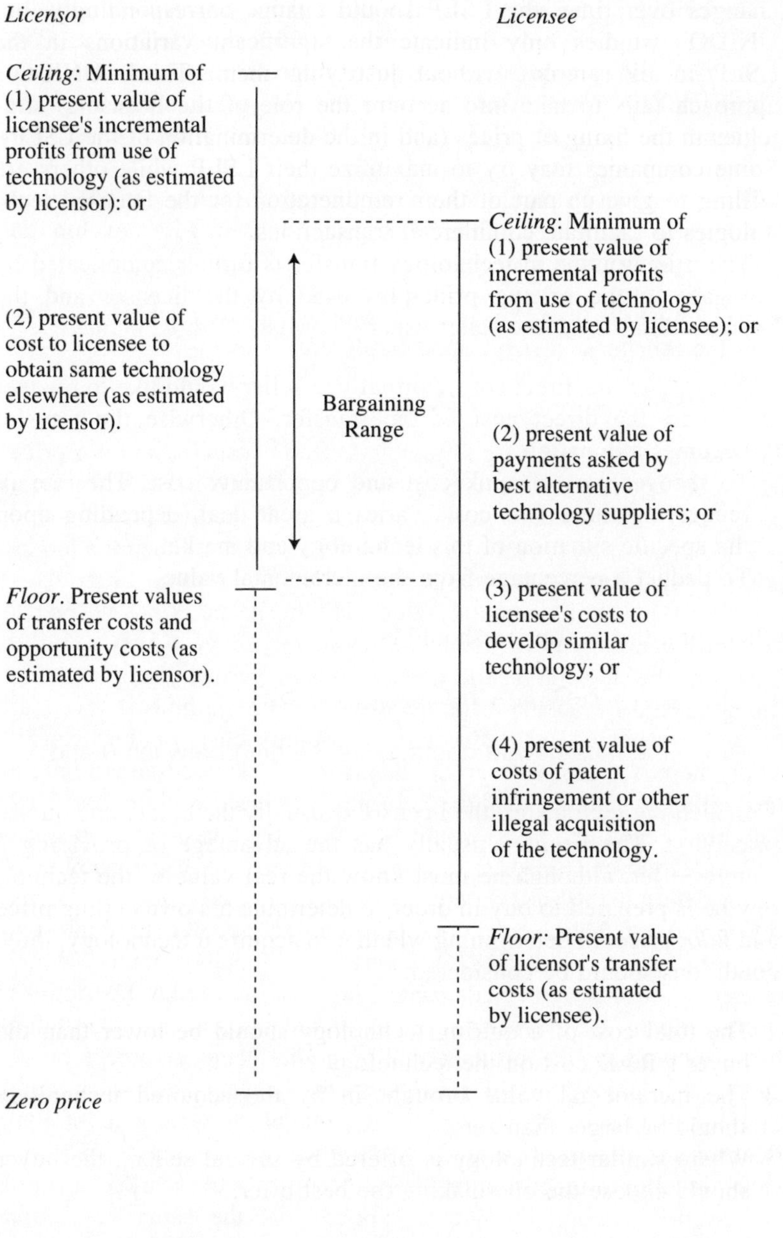

Licensor

Ceiling: Minimum of
(1) present value of licensee's incremental profits from use of technology (as estimated by licensor): or

(2) present value of cost to licensee to obtain same technology elsewhere (as estimated by licensor).

Floor. Present values of transfer costs and opportunity costs (as estimated by licensor).

Zero price

Bargaining Range

Licensee

Ceiling: Minimum of
(1) present value of incremental profits from use of technology (as estimated by licensee); or

(2) present value of payments asked by best alternative technology suppliers; or

(3) present value of licensee's costs to develop similar technology; or

(4) present value of costs of patent infringement or other illegal acquisition of the technology.

Floor: Present value of licensor's transfer costs (as estimated by licensee).

Figure 10.2 The normative model of licensing negotiations
Source: Root and Contractor 1981: 25

1 The ceiling price should not exceed the expected incremental value R. Otherwise, the licensee is deprived of his expected profits from the technology acquisition and loses interest.
2 The ceiling price should not exceed the price offered by the licensor's competitors ($P_{competitor}$). Otherwise, the licensee will go to his competitor.
3 The ceiling price should not exceed the cost of the buyer's R&D on the said technology. Otherwise, the licensee will develop the technology on his own.

When setting the floor price, the licensor would like to at least include both the transfer costs and opportunity costs, i.e.:

$$P_{floor}\text{ (seller)} = C_{direct} + b_2C_{opportunity}$$

The licensor does not normally include R&D costs in the floor price, due to the complexity involved in assigning joint R&D costs incurred over years to a particular agreement in a multiproduct firm.

The licensee, on the other hand, would also like to keep the largest possible share of the economic rents. For that reason, he tends to estimate a lower incremental value. Obviously, he is not willing to buy the technology if the three ceilings set out above are exceeded. In regard to the floor price, the licensee would normally only consider the direct transfer costs of the licensor, because he finds it very hard to get a proper estimation of the opportunity costs of the licensor. Thus, there is a gap between the licensor and the licensee not only at the ceiling price but also at the floor price, as is shown in Figure 10.2. The final negotiated price would lie within the bargaining range.

Besides the gaps in pricing principles, there are also a number of major determinants influencing the final price of a technology licensing agreement. These determinants can broadly be classified into two categories (Cho 1988: 74–75): "agreement-specific factors" which comprise various agreement provisions such as market restrictions, exclusivity of the license, duration of the agreement, quantitative limits on product size, product quality requirements, grantback provisions, tie-in provisions and restrictions on the use of technology; and "environment-specific factors" which include government (of both licensor and licensee) regulations on agreement contents and provisions, level of competition in the licensee's product market, level of competition among alternative suppliers of the same or similar technology, political and business risk in the licensee country, and product and industry licensing norms.

Finally, there is little, if any, price elasticity in technology transfer, because the demand for a technology product, such as process technology, is a function of many factors in addition to price. Beyond a certain point, even a substantial reduction in price is not likely to bring about an increase in the number of licenses granted. Moreover, the licensee is more interested in the performance of the transferred technology rather than the technology itself. When the price for a given technology may constitute merely a small portion of the total revenues at stake for the licensee, other factors, such as the licensor's track record, technical support capability and process efficiencies overshadow the issue of pricing (Watkins 1990: 63–65).

PAYMENT TYPES

The eight payment types described in Table 10.1 can be classified into three groups, the sum constituting the total returns from a licensing agreement (Contractor 1981: 36): one-time fees (front-end or lump sum payments, technical service fees, other service fees); fees relating to licensing activity (total royalties, margins on components supplied or products purchased from licensee); other returns (grant-back values, tax savings, licensee equity).

Front-end or initial lump sum fees are used for several reasons: first, initial fees can guarantee recovery of transfer costs, because royalties are subject to the risks of non-performance; second, initial lump sum fees are necessary when disclosure of proprietary information is involved; third, in countries which place ceilings on royalty rates, increased initial fees allow the licensor to realize targeted returns. On the other hand, the licensee would like payments to be linked to their output to minimize their risks while keeping the licensor continually interested in helping the licensee to produce and sell the products.

When front-end or lump sum payments are adopted, they are paid in instalments throughout a licensing operation. A typical payment schedule offered by the Chinese, for instance, is as follows. The first fee of 5 to 20 per cent of the total contract price is payable shortly after the contract is signed. When paying this fee, the licensee can require the licensor to provide a bank guarantee on the fulfilment of the contract. The second fee of 35 to 50 per cent of the total is payable after the basic engineering package is completed and accepted by the licensee. The third fee of 20 to 30 per cent of the total is payable after the plant starts up. This fee mainly covers the expenses of the licensor in providing technical assistance to plant construction, equipment installation and technical adjustment and the training fee for the personnel of the licensee. The last is a contract fee of 5 to 10 per cent of the total, which can be used as contractual "risk" insurance and is payable after qualified products are manufactured (Li 1989: 129–130).

According to the logic of LSLP, royalties should be based on a percentage of licensee profitability. However, "profits" are often hard to determine due to variations in accounting practice across countries and joint costs in multiproduct operations. There is also the complication of a licensee hiding profit figures from the licensor. Consequently, many licensors are reluctant to base royalties on profits, unless they have good control over management or have good reason to be confident in the profitability of the licensee. At the other end are royalties based on output. But if a large number of products cannot be sold, the licensee may sustain serious losses because the royalty payment is fixed to output throughout the contractual period. Between the two are royalty payments based on net sale value (or net selling price). Many developing countries have required royalties based on net selling price and placed the relevant royalty rate in the range of 2 to 5 per cent.

The determination of royalty rate is also directly related to sales volume. If sales volume is small, the royalty rate should be large; but if sales volume is large, the royalty rate should be small. A common practice to determine royalty rate is the method of analogy, which determines a royalty rate by referring to the royalty rate the licensor has already quoted in the earlier transfer of the same or similar technology or referring to the royalty rate of other companies in transferring similar technology. The licensor is often asked to provide the above rates as a reference for the determination of the rate or a third party is asked to evaluate the rate. According to the statistics of the United Nations Conference on Trade and Development (UNCTAD), the popular royalty rate stands between 0.5 and 10 per cent of the net selling price, with 2 to 6 per cent for most products. Some countries, like China, ask the licensor to hold the royalty rate under 5 per cent (Li 1989: 133). For mass products, such as electronics, the royalty rate is usually placed at less than 3 per cent.

There are two types of royalty rates: fixed and sliding. A fixed royalty rate is one that remains unchanged throughout the whole period of contract implementation. Although it is simple, the fixed royalty rate presents many disadvantages to both sides. If the licensee pays the royalty on the basis of its profits, the licensor will suffer losses when the licensee's profits decrease; if the royalty is paid according to output and sales volume, the gains and losses of the transferee are disconnected. A sliding scale royalty rate changes with sales volume. In the initial period of production, the sales volume is relatively small and the royalty rate should be higher; with an increase of the sales volume, the royalty rate should be correspondingly decreased. For example, a 3 per cent royalty rate is applied to an electronic product when the sales volume stands below 1,000 units; 2.5 per cent is applied when the sales volume increases to 1,001 to 5,000; 1.5 per cent is applied when the sales volume runs to 5,001 to 10,000; and 1.3 per cent is applied when the sales volume is above 10,000 units.

In the practice of international technology transfer, a pure running royalty payment is rare. A more common method of payment is to combine initial fees with running royalties. After signing the contract or the delivery of technological documents, the licensee will pay an agreed sum and then pay running royalties. Initial fees are like a deposit and should at least compensate for the total costs incurred by the licensor in the process of technology transfer, which include the costs of preparing documents, quotations, project design, travel and technical assistance. For the protection of the licensee, initial fees should not be set too high, usually at 10 to 20 per cent of the total technology price while running royalties account for 80 to 90 per cent.

Technical assistance fees cover the costs incurred when the licensor sends technical personnel to the licensee's plant to help install, operate and check equipment. They include wages, travel expenses and allowances of the dispatched technical personnel and management costs of the licensor. In the transactions of international technology transfer, fees for technical and other relevant services are distinct from license fees, negotiated on a cost-plus basis, to be paid by the licensee to the licensor on a per diem or lump sum basis (Contractor 1981: 36).

Finally, price can be significantly influenced by different combinations of payment types as well as other relevant terms. The choice of currency for payment, for example, can also considerably distort negotiated prices. For this reason, the licensor would insist on using a hard currency while the licensee prefers using a soft currency in an international technology transaction. Therefore, licensees tend to carefully compare payment packages from different suppliers. The following example is a simplified version of the impact of payment types on overall pricing:

> A Thai company had decided to acquire technology to manufacture a highly sophisticated telecommunication product. According-

ing to the quotation offered by a German company, the Thai company would pay 4.4 million marks right after the signing of the contract. Then the Thai company would pay the German company the remainder in royalty fees from the end of the third year to the end of the eighth year. The selling price of this product was 100 marks per piece. The planned annual output for this product was 845,000 pieces. The technology recipient's gross profits should be 30% and the supplier's LSLP was set at 16.7%. (1 mark = US$0.628 and i = 10%.) A French quotation was based on a front-end lump sum payment schedule. The total quoted price was 75 million francs. The Thai company would pay a down payment of 20% of the total, then pay 50% by the end of the first year. The remaining 30% would be paid by the end of the second year when the machinery began to produce quality products. (1 franc = US$0.182 and i = 10%.)

Question: Given that the German and French technology is equally good for the Thai company, which quotation is more favorable to the Thai company?

$$G = 4.4 + (16.7\% \times 30\% \times 100 \times 0.845)\frac{(1+10\%)^6 - 1}{10\%(1+10\%)^6}\,\frac{1}{(1+10\%)^2} = 19.60\text{m marks} = \$\,12.32\text{m}$$

$$F = 75 \times 20\% + 75 \times 50\% \frac{1}{(1+10\%)^2} + 75 \times 30\% \frac{1}{(1+10\%)^2} = 67.69\text{m francs} = \$\,12.32\text{m}$$

Answer: If other conditions are similar, the German quotation seems to be much more favorable, because the Thai licensee's risk and financial burden are significantly lower, even though the two quotations do not lead to major present value differences.

SUMMARY AND CONCLUSION

This chapter has discussed revenues and costs of technology transfer, analyzed basic pricing principles and compared payment types. Technology pricing is at the crux of the disputes in transactions of the international transfer of technology between independent licensor and licensee. In order to gain more profit, the licensor often quotes high prices whereas the licensee bargains hard and carefully compares price quotations from different sources. In this sense, the price negotiations of technology transfer are not much different from other types of commercial negotiations, whereby many of the tactics used apply to "win–lose" situations or what is known as a "zero sum" game. The main problem is to create a mutually acceptable formula for the sharing of the incremental value, which is further complicated by the many determinants of technology pricing, discussed earlier in the chapter.

Licensing, however, represents by its nature a "win–win" game. Price negotiations should not be based on the rules of a zero sum game. As technology transfer is characterized by a bilateral monopoly, both parties should follow a cooperative strategy in order to be able to share the economic rent generated by technology transfer. If they cannot reach an agreement based on mutual compromise, this sum of wealth will vanish. Neither of them will benefit from the failure to cooperate (Contractor 1985: 199; Friedman 1990: 285–332). Thus, it is against the fundamental interests of either side to force a hard bargain that will cause resentment later on. A licensing agreement represents a mutual commitment to cooperate for the benefit of both parties. As Root (1981: 129) has commented, "a licensing agreement will remain viable over the longer run only if both parties share in the economic rent created by the technology transfer."

11 Licensing agreement

INTRODUCTION

A licensing agreement refers to the written contract between the licensor and the licensee, which specifies the rights and obligations of each party. Each licensing agreement usually describes the type of licensing agreement, the licensed technology, the licensing restrictions, the obligations of each party and the compensation arrangements (Shahrokhi 1987: 40). International licensing agreements require scrupulous preparation and plenty of flexibility. It is generally agreed that successful cooperation is dependent upon the development of trust between the licensor and licensee. Nevertheless, licensors have good reasons to be cautious about entrusting their intellectual property to overseas companies, especially when the agreement involves core technologies or key markets. Similarly, licensees are unwilling to take the risk of investing their resources if they are worried that they may be replaced by the licensor when the venture generates abundant profits.

Many developing countries have enacted legislation designed to protect local licensees from the superior bargaining position possessed by multinational corporations (MNCs). Much of this legislation has resulted from unpleasant experiences in which MNCs initially entered overseas markets via licensing agreements with local companies. Once the foreign operation became a profitable operation, the licensor replaced the licensee with one of its own subsidiaries. To restrict these practices, host countries often refuse to enforce contractual provisions granting liberal termination rights to the licensor. To promote healthy development of the cross-border licensing business, the United Nations and its organizations have created various model licensing systems, with an emphasis on the interests of developing countries (UN 1973; WIPO 1977; UNECE

1980; and UNIDO 1989). This chapter discusses the key contractual provisions, presents a sample agreement and makes critical comments on the agreement.

KEY CONTRACTUAL PROVISIONS

A licensing venture may encompass separate agreements on patents, trademarks, technical know-how and technical services. For explanatory convenience, the term "licensing agreement" is hereby used to cover all the agreements between the licensor and licensee concerning a particular venture. The precise provisions of a licensing agreement depend on the objectives of the parties and their relative bargaining power, the nature of the intellectual property and the relevant national laws governing the relationship. Therefore, technology licensing agreements may assume different forms. In spite of these differences, however, a number of important provisions should be included in most licensing agreements (Root 1994: 129–132; Richards 1994: 341–344; Ehrbar 1993: 220–241; Sherman 1991: 319–332; Watkins 1990: 95–116; Van Horn 1989: 123–126; Arnold 1984; and Brookhart *et al.* 1980).

Scope of the license

The granting clause describes the precise scope of the license. It addresses the very important question of what exactly is being delivered by the licensor to the licensee in return for payments and fees. In general, it involves a grant by the licensor to the licensee of the right to practice the licensor's technology, normally in connection with the manufacture, use and/or sale of some desired products. It should clearly indicate whether the license is exclusive (where the licensor gives up the right to license the technology to others in the licensee's territory) or non-exclusive. The granting clause may also determine whether the licensee can grant sublicenses.

Occasionally, the granting clause may include a grantback provision that requires the licensee to transfer any inventions or improvements it derives from the licensed technology back to the licensor. For example, when a patent licensee improves on the patented technology covered by the licensing agreement, it is obliged to grant the new intellectual property rights to the licensor. Many developing nations have legislation to prohibit grantback clauses, particularly unilateral grantback clauses, on the ground that they violate local firms' rights of ownership in new technology and strengthen the market domination of foreign licensors. When the agreement encompasses the transfer of trade secrets, many licensees have asked for a technical service clause which requires licensors to provide necessary assistance during the life of the agreement.

Restrictions on use

All licensing agreements include some types of restrictions, with most of them containing some type of territorial restriction. Although the licensee may want to utilize the licensed technology as much as possible in the world, the licensor tends to reserve certain markets for itself or for other licensees by confining the licensee to a specific geographic territory or by imposing quantitative limits on the licensee's use of the technology. Whether the licensor can limit a licensee's right to sell outside its native country also depends on governmental regulations in that country and on laws governing restrictive business practices in the home country.

Licensors may also wish to include other restrictive clauses that preclude licensees from purchasing goods, services or technology from sources other than the licensor, particularly from competing sources. Some of these provisions may restrict the licensee's research and development or require the appointment of technical and management personnel from the licensor. Others may require the acceptance of additional technology not desired by the licensee or impose restrictions on use after expiration of the agreement. Still others may attempt to fix prices. Most of these provisions contain anti-competitive tendencies, and are highly controversial and strictly regulated by national anti-monopoly laws (Finnegan 1980: 38–110). This issue will be discussed in detail in the following chapter.

Performance requirements

The success of a licensing venture depends on the licensee's ability and commitment to manufacture the licensed product up to agreed quality standards and to fully exploit its sales potential in the target market. Besides suffering severe damage to its reputation, the licensor may forfeit its trademark rights in the target country if the licensee fails to meet the normal quality standards. To protect its interests, therefore, the licensor may insist on the inclusion of performance provisions in the agreement concerning both production and sales. Quality control may be executed by the licensor in the following two ways: the right of the licensor to monitor the licensee's operations; and the right of the licensor to use its own engineers to supervise the licensee's production. Likewise, the licensee may insist on warranties to ensure that the intellectual property meets certain performance standards.

Compensation

The licensee's payments for use of the licensed technology may be structured in a variety of ways. Occasionally, the parties may agree on a single lump sum payment due in advance of the technology transfer. Sometimes, front-end lump sum (instalment) payments are preferred. In many cases, a combination of lump sum payment and royalties is used. While the licensor would like the licensee to pay minimum royalties for each contract year, the licensee frequently makes running royalty payments based on a percentage of sales or actual use of the technology. When compensation is tied to actual use, the licensing agreement usually obliges the licensee to keep careful records and/or allows the licensor to inspect the records (Contractor 1985: 173–241 and 1981: 337–354).

Also included are provisions on taxes that must be paid, how they relate to the royalty agreement and who pays them. Customs duties, business taxes, payroll taxes and insurance should be taken into account. Usually, compensation provisions specify the currency in which royalty payments must be made and the way to determine the exchange rate when they are to be made in a currency other than the one in which a licensee's sales were made.

Intellectual property protection and confidentiality

To protect the licensed technology, licensors often clearly state the rights and privileges of the licensee in using it, how and within what scope it should be exploited and what constitutes a misuse. When the licensed technology is a trade secret, the agreement should contain a confidentiality clause that restricts the licensee's right to disclose the information. The licensor may also specify the licensee's responsibility for protecting against infringement by competitors.

There are basically three distinct time frames within which confidentiality provisions are applied. First a licensee may wish to have a limited right to inspect the licensor's technology before accepting the licensing agreement to ensure that the relationship will be valuable. In response, the licensor may require the licensee to sign a secrecy contract prohibiting the latter from exploiting the technology or disclosing it to others. Next, throughout the life of the licensing agreement, the licensor may wish to restrict the number of people with whom the licensee will share its trade secrets. Lastly, the licensor usually requires the licensee to promise not to use or disclose trade secrets after the licensing agreement has expired (Richards 1994: 342).

Dispute settlement

While the provisions of a licensing agreement should be very explicit, they are also required to be sufficiently flexible for both parties to implement them. However, even the best written contract cannot exclude the possibility of disputes from time to time. In most cases, these disputes can be resolved by referring to the written agreement and by negotiating over differences. But with some disputes, settlement may only be achieved by the use of external procedures, which should be clearly stated in the agreement.

Generally speaking, litigation should be avoided by the parties to the agreement, because it is time consuming and costly. For the licensor, a vexing uncertainty is that jurisdiction may be established in the foreign country even if the agreement stipulates that the governing law is that of the licensor's country. One common way to avoid litigation of disputes is to provide for arbitration that is binding on both parties. The arbitration clause in the contract should state the applicable arbitration rules (such as those of the International Chamber of Commerce) or appoint the arbitration tribunal. This is discussed in detail in Chapter 16.

Force majeure

Most parties to an international agreement would like to include this clause in their agreement. *Force majeure* refers to an unforeseeable and irresistible event that renders an obligation impossible to perform. According to the Contracts for International Sale of Goods (CISG) promulgated by the United Nations, a promisor should not be made liable for failure to perform if the non-performance has resulted from an impediment beyond the promisor's control that the promisor could not have been reasonably expected to have considered when the contract was established.

Duration

The length of the technology transfer agreement may depend on the perceived interests and the bargaining position of the parties concerned. Licensees have frequently demanded long-term licenses to ensure they can recover their investments. The duration of the licensing deal is also often subject to the governing national law. Some countries require minimum terms to ensure that local licensees recover their investments, while others impose a ceiling on the maximum term, at the end of which the technology should be awarded to the licensee. Sometimes, the term of the agreement is influenced by the nature of the intellectual property rights. For instance, an agreement transferring rights in patented equipment should not exceed the life of the patent.

Most licensing agreements specify the circumstances under which either party may terminate the relationship. An agreement may be terminated in the event of bankruptcy or judicial or administrative declaration of insolvency of either party. A licensor may revoke the agreement if the licensee fails to reasonably perform, or delays fee or royalty payments or breaches the requirement of secrecy. The license may also be terminated if the licensor fails to provide promised technical services or licenses the technology to another company in the licensee's territory in violation of exclusivity.

A SAMPLE AGREEMENT

The following sample agreement is directly quoted from the case book by Richard D. Robinson (1988a: 95–104).

License agreement

THIS AGREEMENT, effective as of the day of , 1987, by and between INVENTIVE SYSTEMS, INC., a corporation organized and existing under the laws of the State of Ohio, with offices at 570 Parkhurst Avenue, Milford, Ohio (hereinafter referred to as "INVENT"), and ESTRA, A/B, a corporation organized and existing under the laws of Sweden with offices at Storgatan 35, Stockholm, Sweden (hereinafter referred to as ''ESTRA'').

Witnesseth that

WHEREAS, INVENT is the owner of all right, title and interest in and to an invention embodied in the pending US application for patent Serial No. 682,832 entitled "Method of and Electronic Apparatus for Measuring Marine Speeds" and foreign counterparts, and in and to certain trade-secret, know-how and related confidential proprietary information and technology relating to said invention (hereinafter referred to as proprietary technology); and

WHEREAS, ESTRA desires to obtain the right to manufacture, use and sell equipment embodying said invention in Sweden and to obtain such proprietary technology to enable such manufacture, use and sale;

NOW, THEREFORE, in consideration of the premisses, the license granted herein by INVENT to ESTRA and the covenants and conditions herein contained, the parties agree as follows:

Article I

Definitions

Section 1.1 "Licensed Invention" means the invention embodied in pending United States patent application Serial No. 682,832 entitled "Method of and Electronic Apparatus for Measuring Marine Speeds" and/or in any and all continuations and divisions of the same, and any and all improvements upon the same which ESTRA elects to add to this agreement upon prompt notification as to the same by INVENT.

Section 1.2 "Licensed Apparatus" is any apparatus made in accordance with either or both of (1) the disclosure of the United States patent application above identified, and (2) in accordance with said proprietary technology originating with INVENT and disclosed hereunder to ESTRA.

Section 1.3 (a) The term "Sold" includes leased, given, transferred, delivered or retained for use by ESTRA or retained for future sales by ESTRA or as part of inventory to be valued at expiration of this agreement.

(b) The term "Net Selling Price" means:

- (i) when sold in an arm's length transaction for a specified consideration, the price of any Licensed Apparatus less any sales taxes, use taxes, excise taxes and discounts for quantity, and where separately stated, insurance and shipping charges but before deductions of cash discounts, agent's commissions, or any other allowances;
- (ii) in the event that Licensed Apparatus is incorporated as part of a larger equipment and Licensed Apparatus has no separate

price as such, then the price for the purpose of computing Selling Price in (i) above of Licensed Apparatus for royalty purposes shall be computed by taking the ratio of the manufacturing cost of the Licensed Apparatus to the manufacturing cost of the larger equipment (including the Licensed Apparatus) multiplied by the price of the larger equipment;

(iii) upon termination of this agreement ESTRA will direct its Certified Public Accountants to deliver to INVENT an inventory of all Licensed Apparatus and combinations thereof. One year from the termination of this agreement payment pursuant to Section 3.1 hereinafter, shall be due on inventory held by ESTRA at termination and sold subsequent to termination. All inventory still held by ESTRA within the year subsequent to termination shall be priced at then fair market value for final payment pursuant to Section 3.1 herein.

Article II

Grant of license

Section 2.1 Subject to the terms and conditions herein set forth, INVENT hereby grants and agrees to grant to ESTRA for the term of this agreement, a non-exclusive and non-transferable right and license to make, have made, use and sell Licensed Apparatus, to practice the process disclosed in the Licensed Invention and to use said trade-secret, know-how and related confidential proprietary information and technology in the territory including Sweden, only, and shall not sell to another with knowledge of buyer's intent to resell outside of the territory licensed herein.

Section 2.2 At any time during the term of this agreement INVENT shall have the right to obtain a royalty-free, non-exclusive license with sub-licensing rights, to inventions made by ESTRA relating directly to improvements in the Licensed Invention.

Section 2.3 INVENT further agrees to provide ESTRA with engineering support immediately following execution of this agreement for the purpose of transferring said proprietary know-how and technology, and assisting ESTRA in developing the same for its production. Such support will be provided by INVENT to the extent that INVENT's operations will reasonably permit and will be paid for by ESTRA at INVENT's normal engineering rates. Transportation, per diem and other related expenses will be paid for by ESTRA. All payments made by ESTRA for engineering and support services, expenses and related items, are exclusive of the royalties to be paid under Sections 3 and 4. INVENT is deemed to have fully performed its obligation under this agreement with the completion of six (6) months of engineering service.

Article III

Royalties payable by

Section 3.1 ESTRA shall pay to INVENT, for the life of the last to expire patent licensed hereunder, the following percentage royalties during the term of this agreement, based upon the net selling price of each Licensed Apparatus sold by ESTRA.
(Still subject to negotiation.)

Section 3.2 In the event that, commencing after the fifth anniversary of this agreement, the Licensed Apparatus thereafter manufactured and sold by ESTRA hereunder shall not be covered by one or more claims of an issued patent of INVENT, all such royalties provided for in Section 3.1 of this Article shall be payable only for period of ten years from the date of this agreement.

Section 3.3 Only one royalty shall be payable hereunder in respect to any Licensed Apparatus, and such royalty shall be payable with respect to the first use of sale thereof.

Section 3.4 Licensed Apparatus shall be considered sold, except with respect to Section 1.3 when billed out, or if not billed out, when delivered; or when paid for if paid for before delivery. Royalty payments on Licensed Apparatus not accepted or returned to ESTRA for credit by its customers shall be credited on future royalty payments.

Section 3.5 Irrespective of actual sales hereunder, ESTRA agrees to pay INVENT (amount under negotiation) as non-returnable down-payment upon the execution of this agreement; and thereafter, for the life of this agreement, a minimum annual royalty of (not agreed upon), payable in quarterly instalments on or before the last day of each January, April, July and October together with its accounting pursuant to Section 4.1 hereinafter set forth in Article IV hereof, and creditable against the year in which due, only.

Article IV

General provisions relating to royalties

Section 4.1 ESTRA shall keep full and accurate records of all Licensed Apparatus manufactured and sold or otherwise disposed of by ESTRA. On or before the last day of each January, April, July and October, during the term of this agreement, ESTRA's Certified Public Accountants shall render a statement to INVENT setting forth the type and total number of Licensed Apparatus sold by ESTRA during the preceding calendar quarter and the amount of royalties payable hereunder on account thereof. Within thirty (30) days after the termination of this agreement, ESTRA will render such a statement to INVENT covering: any period prior to termination for which a statement had not been previously rendered together with its report pursuant to Section 1.3. Each such statement shall be accompanied by payment to INVENT of the royalties shown thereby to be due.

Section 4.2 ESTRA shall pay interest to INVENT upon any and all amounts of royalties that are at any time overdue and payable to INVENT at an annual rate of two per cent over prime during the time commencing thirty (30) days from the last date of the period covered by such royalties as provided herein, and continuing to the date of payment.

Section 4.3 ESTRA agrees to permit the aforesaid records to be examined at all reasonable times during normal business hours by a Certified Public Accountant retained by INVENT and acceptable to ESTRA.

Article V

Term and termination

Section 5.1 This agreement and all rights granted hereunder shall extend for the respective full lives of any patents which may be issued on any invention herein licensed, including improvements added to this agreement by INVENT unless terminated earlier in accordance with Sections 5.2 or 5.3 of this Article, or Section 3.2 of Article III hereof.

Section 5.2 ESTRA may terminate this agreement upon the sixth anniversary thereof by giving INVENT written notice of its intent so to terminate at least sixty (60) days prior to said anniversary date. ESTRA shall also have the right, upon giving the same notice prior to each subsequent five-year interval, to terminate the license herein granted, and unless so notified, this agreement shall continue in full force and effect for additional five-year terms.

Section 5.3 If ESTRA shall at any time default in making any payment required under this agreement or in making any report hereunder, or commit any breach of any covenant or agreement herein contained, and shall fail to remedy any such default or breach or to correct any such report within sixty (60) days after written notice by INVENT, INVENT may, by notice in writing to this effect, and at its option, terminate this agreement and the license granted hereunder, but such termination shall not prejudice INVENT's rights which shall have accrued to the date of such termination, nor its rights to recover any royalty or any sum due at the time of such termination nor shall it prejudice any cause of action or claims of INVENT accrued or to accrue on account of any breach or default by ESTRA. Any waiver by INVENT of any of its rights under this agreement shall not be deemed a waiver of any other rights under this agreement, nor shall any waiver by INVENT of any breach of this agreement be deemed a waiver of any other or subsequent breach.

Article VI

General provision

Section 6.1 The parties contemplate that certain of said INVENT technology will be imparted to ESTRA during the term of this agreement, and it is agreed that ESTRA will maintain all such identified by INVENT as confidential to the same degree secrecy as it applies to its own proprietary information for a period of five (5) years beyond the termination of this agreement as spelled out in Section V, provided, however, that ESTRA is not obligated by this agreement to maintain in confidence such information which shall have been in its possession prior to its delivery to ESTRA or is in the public domain or which may become known to ESTRA independently of INVENT.

Section 6.2 ESTRA shall not have the right to assign its rights under this agreement without the prior written consent of INVENT, except that ESTRA shall have such right, at any time, without INVENT's consent, as a result of sale of a division manufacturing the Licensed Apparatus, merger or consolidation. Any such assignee shall have all rights of ESTRA hereunder as if this agreement had originally been made between INVENT and such assignee. However, rights herein granted shall in no way be construed so as to permit transfer of said license as restricted in Section 2.1 hereof.

IN WITNESS THEREOF, the parties hereto have caused two (2) copies of this agreement to be executed by their duly authorized officers and their corporate seals to be hereunto affixed as of the date specified in the heading of this agreement.

(Seal) INVENTIVE SYSTEMS, INC.

Attest: by

____________________ ________________________

Secretary President

(Seal) ESTRA, A/B

Attest: by

____________________ ________________________

Secretary

CRITICAL ANALYSIS OF THE AGREEMENT

The licensing agreement between INVENT and ESTRA, simple as it may seem, contains most of the key provisions described in the first section of this chapter. It is almost a pure license, though Inventive Systems does undertake to provide engineering support (section 2.3). There is also a separate letter on confidential disclosure. The technology concerned is both a process and product technology. It can be characterized as an improvement technology, because it involves no really new technology, but only a unique application. Although the technology is fairly complex, it does not seem to require skills not readily available in the market. The technology can be reverse engineered by reasonably skilled electronic engineers.

One major deficiency of this agreement is the absence of a dispute settlement mechanism. There are no adjustment clauses against the unpredictable changes in the condition of business. *Force majeure*, which is an unforeseeable and irresistible event that renders an obligation impossible to perform, is conspicuously missing. Most parties to international contracts will include a *force majeure* clause in their private agreement. There is also no choice of law and forum selection clauses in the event of a dispute. Neither is there an arbitration clause. If there is a bitter dispute in the future, the chance of a private settlement may be significantly reduced. When litigation becomes necessary, it may be more time consuming and costly.

Another major deficiency is that there are no provisions specifying trade names or trademarks to be used. If trade names and trademarks are introduced, will they be exclusive to ESTRA, or will INVENT also have the right to use them? This can cause tremendous problems to the two parties, especially for the licensor. If the licensee gains great success in the market by using its own trade names and trademarks, the licensor may have great difficulty in penetrating that market.

Finally, one should notice that the level of royalty and of the minimum annual royalty have not been agreed upon. This is usually one of the most difficult parts of negotiation. Many negotiations fail on this issue. This means that there may still be lengthy negotiations before both parties eventually reach a mutually satisfactory agreement (Bidault 1989).

From the viewpoint of the licensor, there may be several reservations about signing the agreement as it now stands, in addition to the above three points (Robinson 1988b: 13). First, the scope of the technology as defined seems to be too vague (section 1.2). Second, if ESTRA sets up a sales subsidiary and sells at heavily discounted prices, the licensor may sustain losses in revenue. Third, while the agreement specifies the need for INVENT to monitor ESTRA's manufacturing costs, there is no mechanism to guarantee that INVENT has access to an accurate accounting record of ESTRA's sales, and the monitoring operation may be both difficult and costly. Fourth, up-front lump sum payment for the initial transfer is not specified.

From the licensee's viewpoint, the grants of the licensor do not seem really generous, as it grants only "a non-exclusive and non-transferable right and license" of the said technology to the licensee within Sweden. While a non-returnable downpayment and minimum royalty are stipulated under the royalty section, there are no warranties whatsoever in the agreement to guarantee that the intellectual property meets certain performance standards. The paid engineering support "is deemed to have fully performed its obligation under this agreement with the completion of 6 (six) months of engineering service." If the licensee fails to grasp the transferred technology and know-how within the time limit, it may have to take the full responsibility and consequences. Finally, the commitment to grant a "royalty-free, non-exclusive license with sublicensing rights, to inventions made by ESTRA relating directly to improvements in the licensed invention," resembles a free grantback unfavorable to the licensee.

A good licensing agreement usually is not sufficient to ensure the success of technology transfer. Numerous operational and administrative support activities after the execution of the agreement constitute the keys to success or failure (Watkins 1990: 11–127). Operational support involves such activities as the preparation of process packages, review of detailed design drawings and specifications, and assistance in the start-up phase. Administrative support is designed to ensure contract compliance, organizational continuity, and an efficient technology transfer interface for outside contacts. One key element in administrative support lies in a good data base containing abstracts of all licensing and relevant agreements, with a focus on contractual deadlines and commitments. A carefully prepared licensing support program can ensure that both the parties will turn out to be the winners with their expectations fulfilled.

SUMMARY AND CONCLUSION

This chapter has discussed the key provisions of international licensing agreements, introduced a sample agreement for deliberation and made a critical analysis of the sample agreement. As has already been emphasized, a licensing agreement is the key to the success of a licensing business deal. Although terms may vary a great deal in different licensing agreements, some basic provisions can be found in most agreements. When a licensing agreement is negotiated between parties from different countries, it should also include clauses on the choice of language, choice of currency, choice of laws, forum selection and arbitration. Meanwhile, both sides should make appropriate efforts to build up trust for a relatively long-term relationship in the years to come.

In addition, both licensor and licensee should take due precautions to ensure that the intended provisions of their agreement are in conformity with the policies and laws of the country where the technology is licensed. The host government, for example, may wish to restrict agreements that drain hard currency from the country. Thus, when the government finds that the agreement contains provisions draining hard currency, it may not grant approval, or it may significantly delay approval or even interfere in the final stages in the implementation of the agreement. If the government's policy poses a major barrier, close cooperation between the licensor and licensee is necessary to convince the government that the benefits will outweigh the costs. Among the main issues involving governmental interference is restrictive business practice, which will be discussed in the following chapter.

12 The issue of restrictive practices in technology transfer

INTRODUCTION

One outstanding feature of technology transfer is that both sides of the deal are involved in similar kinds of business and produce similar kinds of products. Between them, there exists the possibility of both cooperation and conflicts of interests. In the process of technology transfer, licensors tend to impose several restrictive terms on licensees by utilizing the advantage of their protected industrial property. In order that licensees will accept these restrictive provisions, they tend to interpret them as part of common international trade practices. While licensors do have valid concerns on their transfer of technology, many of these imposed restrictions fall into the category of restrictive business practices.

Restrictive business practices are a common international problem. In order to protect their own interests, licensees are always opposed to this kind of monopolistic practice. Many technology transfer deals cannot be accomplished due to the attempt of licensors to impose certain restrictive terms. Therefore, restrictive practice has become a major obstacle to the development of international technology transfer. To encourage fair competition and protect fair business interests, most countries in the world have enacted laws to control restrictive business practice. The United Nations has attempted to create international rules to solve this problem, including a technology transfer draft code (UNCTAD 1985). The World Intellectual Property Organization has also prepared a model licensing guide for developing countries (WIPO 1977). But so far, there are no unified international rules. This chapter discusses the nature and content of restrictive business practices, reviews anti-restrictive business practice laws in both developed and developing countries and examine the model laws of the United Nations, in an attempt to offer a basic introduction to this complicated international issue.

THE NATURE AND CONTENT OF RESTRICTIVE BUSINESS PRACTICES

Definitions

Restrictive business practices are variously defined. Most of the definitions pertain to questions of market dominance by one firm or collusion involving two or more firms, and the consequences for competition. The concept is therefore associated with market asymmetry and is related to the study of imperfect rather than perfect markets (Long 1981: 1). As has been mentioned earlier, the technology market is indeed an imperfect market, whereby monopolistic practices of technology suppliers or bilateral monopolies of both suppliers and recipients are not uncommon.

For the purpose of this chapter, the definition adopted in the famous *Havana Charter, Final Act,* by the United Nations Conference on Trade and Development (United Nations 1948) still seems applicable: restrictive business practices refer to attempts by firms to "restrain competition, limit access to markets, or foster monopolistic control." In international technology transfer, restrictive business practices are regarded as attempts by suppliers to restrict the market entry of recipients or fair competition by taking advantage of their monopolistic superiority in technology and dominant position in the market.

Parties to a technology transfer deal have both cooperative relationships and conflicts of interests. Their cooperation may lead to their joint attempt to monopolize the market, while their conflicts of interests may induce one side to impose restrictions on the other. Therefore, the nuclear issue of restrictive business practices is monopoly. However, one must make a clear distinction between dominant position in a market and the abuse of this position. The former is legal while the latter is illegal. One prominent feature of the contemporary patent system is to grant the owner of a patent protection in the form of monopoly. If the patent holder abuses this monopolistic position (or dominant position) by imposing various irrational additional conditions, it commits restrictive business practices, which are not permissible under most national laws.

Major items of restrictive business practices

There are many kinds of restrictive business practices, which vary a great deal in terms of the scope and degree of restriction. But, in technology transfer deals, major restrictive business practices often include restrictions first on the use of technology, and second on production inputs and the sale of products (Blakeney 1989: 35–42; Finnegan 1980: 38–110; and Cabanellas, Jr 1984: 51–156).

The use of technology

This category includes the following types of restriction:

1 Use of competing technology or tie-out clauses. Other than for reasons of protecting confidentiality, there is very little justification for this category of restraint. From the viewpoint of a technology recipient, a prohibition on the acquisition of competing technologies will inhibit the very important job of comparative assessment of technologies in order to determine their appropriateness to the circumstances of the recipient.
2 Restrictions on research and development. The licensee is prohibited from conducting further research on, improvements of and adaptations to the licensed technology. Such provisions are widely criticized as impermissible restraints.
3 Grantback provisions. Such provisions oblige a licensee to transfer to the licensor any improvements in the technology embodied in the licensed industrial property rights. These provisions are viewed with antagonism where they are imposed without the reciprocal obligation of the licensor to supply improvements to the licensee, because they strengthen the dominant position of the licensor and tend to stifle the incentive of licensees to conduct adaptive R&D.
4 Package licensing. This requires the acceptance of additional technology not desired by the recipient, as a condition for obtaining the technology in question, and requires remuneration for such additional technology. While this practice should be prohibited on most occasions, the prohibition should not inhibit package licensing where the elements of the package can be shown to be indispensable.
5 Restrictions on use after expiration of the agreement. Where the licensed industrial property right is still in existence at the end of the license agreement, most countries tend to respect those restrictions which derive from the rights granted by the relevant industrial property. However, without valid property rights, the supplier

has no legitimate claim to royalties or control on use after expiration of the agreement.

6 Unused technology. This refers to provisions which require payment or continuation of payment for unused or unexploited technology. Such provisions are opposed by many countries, especially when the technology has not been sought.

Production inputs, and production and the sale of products

This category includes the following types of restriction:

1 Tying (or tie-in). Technology is licensed on the condition that the recipient will purchase additional inputs such as plant and machinery, raw materials, intermediate products and additional technology from the supplier or a source designated by it. Although the imposition of a tie may be justified as conducive to the successful operation of the technology, the opportunity of using an industrial property monopoly as a basis for the tie-in of other items is not an uncommon abuse of industrial property rights.
2 Appointment of technical and management personnel. Closely related to the tie-in of goods and services to the supply of licensed technology is the obligation of a licensee to use personnel designated by the supplier. Personnel restraints tend to be viewed with suspicion by the recipient, though permission is often granted on the condition that the licensor undertake the training of local personnel to replace those supplied by the licensor.
3 Exports. Export restrictions are very unpopular in developing countries, as one major reason for these countries to seek new technologies is to improve the balance of trade and foreign exchange holdings. However, the licensor may have a legitimate concern in preserving its existing overseas markets and protecting them from competition by its licensees.
4 Price fixing. This refers to provisions and/or practices whereby the supplier of technology reserves the right to fix the sale or resale price of the products manufactured. Such action has been regarded as illegal under most antitrust laws.

Principal differences between developing and developed countries

Restrictive business practice is a very difficult issue dividing not only the suppliers and recipients of transferred technology but also their home countries. Historically, developing countries have adopted a broad definition of restrictive business practices and a very tough stance against them while developed countries have been fairly limited in their interpretations of restrictive business practices and quite reserved in handling such cases. With the collapse of the Soviet bloc and market reform in many socialist or former socialist and developing countries, a number of these countries have begun to re-examine their traditional approach toward restrictive business practices in the context of questing for Western technology and investment. A much clearer distinction is made between legitimate and illegitimate restrictive business practices.

According to Li (1993: 144–146), in dealing with restrictive business practices involving technology transfer, one must not only stand firm but also be capable of being flexible. On some restrictive practices, such as tie-in clauses, restrictions on research and development, restrictions on use of competing technology or tie-out clauses, or unused technology that has not been sought by the recipient, the recipient party should stand firm and resist the attempt of the supplying party. These restrictive clauses have exceeded the scope of legitimate patent protection. Both the antitrust laws of developed countries and the technology transfer laws of developing countries (to be discussed in the following sections) can be used to inhibit such irrational conditions.

On other restrictive business practices, such as grantback clauses, restrictions on volume, restrictions on export and restrictions on use after expiration of the agreement, the recipient party should take a flexible stance. On many occasions, the supplying party does have a valid reason to impose these restrictive clauses. If the recipient party resists such clauses without analyzing specific situations, it may lose a good opportunity to receive the technology that it badly needs. However, even though the supplying party has valid grounds, careful negotiations on these restrictive clauses are necessary to ensure that the legitimate rights of the recipient party are not violated.

LAWS REGULATING RESTRICTIVE BUSINESS PRACTICES IN DEVELOPED COUNTRIES

Since the beginning of the twentieth century, most developed countries have enacted laws to regulate restrictive business practices. Although these laws vary a great deal in terms of content and name, they have been enacted to protect free trade, encourage fair competition and control monopoly. Among the most typical are the antitrust laws of the United States, Japan and the European Union (formerly the European Economic Community).

United States

Attempts to exercise control over restrictive business practices in the United States date back to the Sherman Act of 1890. Other major laws in this area include the Clayton Act (1914), the Federal Trade Commission Act (1914), the Webb Pomerene Act (1918) and the Robinson–Patman Act (1936). The most significant antitrust law is still the Sherman Act, section 1. The Clayton Act was enacted to tackle specific restraints of trade which Congress felt had not been adequately prohibited by the courts under the broad mandate of section 1 of the Sherman Act. Section 3 of the Clayton Act, for example, prohibits certain tying arrangements and exclusive dealing contracts involving sales and leases of goods (Marks 1984: 157–178; International Patent and Know-how Licensing Task Force 1981).

Penalties against antitrust violations in the United States are among the most severe. Under the Sherman Act, penalties for individuals include fines up to $350,000 per violation and up to three years in prison; corporations may be fined up to $10 million per violation. Section 4 of the Clayton Act permits any individual who suffers antitrust damage in his or her “business or property” to bring a private civil action against the offender. Successful plaintiffs may recover treble damages, plus reasonable costs and attorneys’ fees (Cheeseman 1992: 1066).

Nevertheless, exemptions from the prohibition on concerted action contained in the Sherman Act are made by the Webb Pomerene Act for firms engaged in export trade. Considerable leeway is allowed for firms’ national laws because some are involved in international production to avoid regulation problems pertaining to the extraterritorial application of national laws. This has remained so even though some cases, such as Union Carbide in 1962, have indicated that action has been taken against such firms. The main criterion for legal action against firms operating internationally is whether or not they have a substantial impact on American foreign commerce. To put it more clearly, national interests are the supreme concern (Long 1981: 32).

Japan

As early as 1947, Japan had its anti-monopoly and fair trade law, though many have argued that Japan has until most recently not executed it vigorously. It is meant to prohibit private monopoly, unreasonable restraint of trade and unfair business practices, by preventing the excessive concentration of economic power and by eliminating unreasonable restraint of production, sale, price, technology and all other improper restriction of business activities through combinations, agreements, etc. This law was designed to promote free and fair competition, to stimulate business initiatives, to encourage business activities and to raise the level of employment and people's income, and thereby to advance the democratic and healthy development of the national economy and to protect the interests of consumers (Hirose 1984: 109).

In 1968, the Fair Trade Commission (FTC) first announced guidelines on patent and know-how licensing agreements concluded between foreign and Japanese enterprises. In the two decades after the announcement of these guidelines, the circumstances concerning technology transfer between Japanese and foreign enterprises have changed dramatically. In the earlier days, Japanese enterprises were primarily recipients of technology from abroad; but they were later on in the position to license technology to foreign enterprises in addition to receiving it. In light of this change, the FTC drafted a new set of guidelines, which was published on February 15, 1989. The guidelines are entitled "Guidelines on the Regulation of Unfair Business Practices in Patent and Know-How Licensing Agreements." Under the new guidelines, the following items may be held unlawful (Matsushita *et al.* 1989: 195–197):

1 To prohibit, during the term of the license agreement, the licensee from handling competing products or adopting the competing technology.
2 To impose on the licensee the condition that the licensee sell the patented product through the licensor or a party designated by the licensor or does not sell to a party designated by the licensor.
3 To impose on the licensee the obligation to inform the licensor of knowledge or experiences which the licensee has acquired with respect to the licensed technology or to grant to the licensor a non-exclusive license of the improvement inventions or applied inventions which the licensee has accomplished, except for cases in which the licensor's obligations are similar in kind and the obligations of both of the parties are balanced.
4 To impose on the licensee an obligation to use the trademark or other representations designated by the licensor on the patented product.

5 To set the standard for the quality of raw materials, components, or the patented product, except for cases in which such a restriction is necessary to maintain the goodwill of the trademark or the usefulness of the licensed technology.
6 To impose on the licensee an obligation to purchase raw materials or parts from the licensor or a party designated by the licensor, except for cases in which the maintenance of the goodwill of the trademark or the effective use of the licensed technology without restrictions on the quality of raw materials or components is difficult.
7 To restrict the areas to which the licensee can export, except for cases in which the restricted areas are: (a) where the licensor has registered the patent, (b) where the licensor is continuously engaged in sales activities of the patented product, or (c) where the licensor has granted the exclusive right to sell to a third party.
8 To set the export price or quantity of the licensee or to obligate the licensee to export through the licensor or the party designated by the licensor, except for the cases in which the export area falls under (a), (b) or (c) listed above and in which the restriction or obligation is reasonable.
9 To require the licensee to pay royalties on products other than the patented product, except for cases in which the quantity of manufacture/sale of the finished products or the value of manufacture/sale thereof is used for the sake of convenience as the basis for computing royalties when the licensed patent relates to a part of the manufacturing process or to components or in which the quantity or frequency of the use of raw materials or components necessary to manufacture the patented product is used as the basis for computing royalties.
10 To require the licensee to accept a license of a plural number of patents en bloc, except for cases in which such a restriction is imposed to the extent necessary to guarantee the usefulness of the licensed patent.
11 To impose on the licensee a unilateral disadvantage in cancelling a license agreement, such as to stipulate that the agreement is subject to immediate and unilateral cancellation, without providing for an adequate grace period, for reasons other than the inability to perform (the agreement) on the part of the licensee due to insolvency or other justifiable causes.
12 To prohibit the licensee from contesting the validity of the patent.

European Union (EU) and its members

The European Union's efforts to control restrictive business practices within the Union have been aimed at setting a code of business conduct for enterprises operating in the regional market of the community. The Treaty of Rome (1957) was designed to serve the wider competitive interests of the Common Market or European Economic Community (EEC) (the predecessor of the EU) as a whole. The provisions cover the EEC's multinational corporations (MNCs) operating within the Common Market, as well as other MNCs, most of which are Japanese and American firms operating in Western Europe (Long 1981: 39–41).

Article 85 of the Treaty deals with restrictive business practices that arise as a result of collusive and other agreements between firms, and Article 86 is related to the abuse of dominant market power by firms. Article 85 prohibits "any agreement between enterprises, any decision by associations of enterprises, and any concerted practice which is likely to affect trade between member states and which has as its object or result the prevention, restriction, or distribution of competition within the Common Market." These include the following items:

1 Price fixing or any other trading condition.
2 Restriction or control of production, markets, technical development or investment.
3 Market sharing or sharing of sources of supply.
4 Differential treatment that leads to competitive disadvantages.
5 Tied purchases that by their nature or commercial usage have no connection with the subject of the contract.

Article 86 includes the following prohibited restrictive business practices, which are associated with market practices by dominant firms:

1 The direct or indirect imposition of any inequitable purchase or selling prices or of any other inequitable trading conditions.
2 The limitation of production, markets or technical development to the prejudice of consumers.
3 The application to parties to transactions of unequal terms in respect to equivalent supplies, thereby placing them at a competitive disadvantage.
4 The subjecting of the conclusion of a contract to a party's acceptance of additional supplies that, either by their nature or according

to commercial usage, have no connection with the subject of such contract.

Articles 85 and 86 symbolize the major progress made by the European Union toward guaranteeing the competitive process outside national boundaries of individual member nations. Nevertheless, the control of restrictive business practices under Articles 85 and 86 does not apply to commercial transactions affecting third countries outside the European Union. There are many cases in which MNCs from the EU member nations have committed restrictive business practices outside the European Union, even though such practices are regarded as illegal within the European Union itself.

As has been mentioned above, member nations of the European Union also have their own rules and regulations. In Britain, for example, some of the major restrictive business practice laws that have been enacted include the Monopolies and Restrictive Business Practices Act of 1948, the Restrictive Trade Practices Act of 1968 and the Fair Trading Act of 1973. In Germany, the main existing law for the control of restrictive business practices can be traced back to the Act against Restraints of Competition as amended in 1957. Some of the major laws which have been in force in France for the control of restrictive business practices can be found in The Amended Decree of August 1959 and Ordinance of September 1967.

LAWS REGULATING RESTRICTIVE BUSINESS PRACTICES IN DEVELOPING COUNTRIES

Since the early 1960s, developing countries have expanded their importation of technology and equipment from developed countries in order to expedite their economic development. In order to protect their burgeoning industries from monopolistic business practices by Western multinational corporations, many governments have enacted laws of technology transfer and established special institutions to supervise and monitor the process of transfer. In many developing countries, international technology transfer contracts cannot take effect until the institutions concerned in the relevant countries have approved. The Andean Pact and Chinese principles are hereby used as examples.

Andean Pact (1971)

Under Article 20 of the Common Regime of Treatment of Foreign Capital and of Trademarks, Patents, Licenses, and Royalties approved in 1971, member countries are not permitted to authorize the conclusion of contracts for the transfer of foreign technology or patents which contain:

1 Clauses by virtue of which the furnishing of technology imposes the obligation for the recipient country or enterprise to acquire from a specific source capital goods, intermediate products, raw materials and other technologies, or of permanently employing personnel indicated by the enterprise which supplies the technology. In exceptional cases, the recipient country may accept clauses of this nature for the acquisition of capital goods, intermediate products or raw materials, provided that their price corresponds to current levels in the international markets.
2 Clauses pursuant to which the enterprise selling the technology reserves the right to fix the sale or resale prices of the products manufactured on the basis of the technology.
3 Clauses that contain restrictions regarding the volume and structure of production.
4 Clauses that prohibit the use of competitive technologies.
5 Clauses that establish a full or partial purchase option in favor of the supplier of the technology.
6 Clauses that obligate the purchaser of technology to transfer to the supplier the inventions or improvements that may be obtained through the use of the technology.
7 Clauses that require payment of royalties to the owners of patents for patents which are not used.
8 Other clauses with equivalent effects.

China

According to the Regulations of the People's Republic of China on Technology Import Contract Administration, promulgated on May 24, 1985, the supplier of technology must not force the recipient to accept unfair restrictive requirements. According to the law, provisions which restrict in any of the following ways must not appear in the contract unless specially permitted by the approving authority (Goosen 1987: 161–162):

1 Additional conditions which are irrelevant to the technology to be acquired by the recipient, including purchase of unnecessary technology, technical service, raw materials, equipment or products.
2 Conditions which restrict the recipient from making free choices

for the procurement of raw materials, spare parts or equipment from other sources.

3 Conditions which restrict the recipient from further developing or improving the technology to be acquired.
4 Conditions which restrict the recipient from obtaining from other sources similar technology or competitive technology of the same kind.
5 Unequal conditions between the two contracting parties for exchanging technical improvements.
6 Conditions which restrict the quantity, variety or sale price of products the recipient may turn out by applying the technology to be acquired.
7 Irrational restrictions on the sales channels and export markets of the recipient.
8 Conditions which forbid the continued utilization of the acquired technology by the recipient on expiration of the contract.
9 Conditions which require the recipient to pay for or be under obligation in respect of unused or invalid patents.

All the major developing countries, such as India, Brazil and Mexico, have similar principles against restrictive business practices, though they vary in degree and scope. It seems that with the liberalization of their economies in the late 1980s and early 1990s, they have adopted increasingly flexible approaches toward restrictive practices. In January 1990, for example, Mexico published new technology transfer regulations applying to the licensing law enacted in 1982. The new regulations have adopted a much more amenable approach toward foreign technology and created a significantly more favorable environment for the transfer of technology in Mexico by multinational corporations.

UN INTERNATIONAL CODE OF CONDUCT ON THE TRANSFER OF TECHNOLOGY

The United Nations Conference on Trade and Development (UNCTAD) has played a key role in creating an international code of conduct on the transfer of technology. In 1970, UNCTAD established the Intergovernmental Group on the Transfer of Technology. In 1974, it received for consideration a draft code of conduct on the transfer of technology which had been prepared by a group of experts under the auspices of the Pugwash Conference on Science and World Affairs. The Group of Experts prepared a Draft Code for a Negotiating Conference of the United Nations. However, the Negotiating Conference, which met in October and November 1978, was unable to adopt an agreed text. Further diplomatic rounds in 1979, 1980, 1983, 1985 and 1987 settled most issues in the code, but have been unable to resolve differences over the content of the restrictive business practice provisions of the code, the applicable law and the settlement of disputes (Roffe 1985: 689–707; and UNCTAD 1987: para 2).

In its most recently published form, the Draft Transfer of Technology (TOT) Code comprises a preamble and nine chapters covering definitions and scope of application, objectives and principles, national regulation of transfer of technology transactions, restrictive business practices, responsibilities and obligations of parties to transfer of technology transactions, special treatment for developing countries, international collaboration, international institutional machinery and applicable law and settlement of disputes (UNCTAD 1985).

Objectives and principles of the TOT Code

The objectives of the TOT Code are "to establish general and equitable standards on which to base the relationships among parties to transfer of technology transactions and governments concerned, taking into consideration their legitimate interests, and giving due recognition to special needs of developing countries for the fulfillment of their economic and social development objectives" (UNCTAD 1985). It promotes international transfer of technology transactions, particularly those involving developing countries, under conditions where the bargaining positions of the parties to the transactions are balanced so as to avoid abuses of a stronger position, so that both sides will be able to accomplish mutually satisfactory agreements.

The main principles articulated in the TOT Code are:

1 All countries should facilitate and increase access to technology, particularly for developing countries.
2 A technology-supplying party should respect the sovereignty and laws of the receiving party and act with proper regard for that country's declared development policies and priorities and endeavor.
3 Special treatment in the transfer of technology should be granted to developing countries.

Unlawful practices

The TOT Code excludes restrictive practices from technology transfer contracts and specifically describes twenty restrictive practices (UNCTAD 1985):

1. Provisions that require the acquiring party to transfer or grant back to the supplying party, or to any other enterprise designated by the supplying party, improvements arising from the acquired technology.
2. Provisions that require the acquiring party to refrain from challenging the validity of patents and other types of protection for inventions involving the transfer or the validity of other grants claimed or obtained by the supplying party.
3. Restrictions on the freedom of the acquiring party to enter into sales, representation or manufacturing agreements relating to similar or competing technologies or products.
4. Provisions that restrict the acquiring party either in undertaking research and development directed at absorbing and adapting the transferred technology to local conditions or in initial R&D programs in connection with new products, processes or equipment.
5. Restrictions on the use of personnel that unreasonably require the acquiring party to use personnel designated by the supplying party.
6. Provisions that unjustifiably impose regulation of prices to be charged by acquiring parties in the relevant market to which the technology was transferred for products manufactured or services produced using the technology supplied.
7. Restrictions on adaption that unreasonably prevent the acquiring party from adapting the imported technology to local conditions or introducing innovations in it, or oblige the acquiring party to introduce unwanted or unnecessary design or specification changes.
8. Agreements that require the acquiring party to grant exclusive sales or representation rights to the supplying party or any person designated by the supplying party, except as to subcontracting or manufacturing agreements wherein the parties have agreed that all or part of the production under the technology transfer arrangement will be distributed by the supplying party or any person designated by him.
9. Tying arrangements that unduly impose acceptance of additional technology, future inventions and improvements, goods or ser-

vices not wanted by the acquiring party or unduly restrict sources of technology, goods or services.

10 Unjustified restrictions that prevent or substantially hinder export, by means of territorial or quantitative limitations or prior approval for export or export prices of products or increased rates of payments for exportable products resulting from the technology supplied.
11 Patent pool or cross-licensing arrangements among technology suppliers and other arrangements that place restrictions on territories, quantities, prices, customers or markets, which unduly limit access to new technological development or which would result in an abusive domination of an industry or market with adverse effect on the transfer of technology.
12 Restrictions that unreasonably regulate the advertising or publicity by the acquiring party.
13 Payments and other obligations after expiration of industrial property rights.
14 Restrictions on the use by the acquiring party after expiration of the arrangement of the technology, including know-how, which has lost its secret character independently of the acquiring party.
15 Limitations on the scope, volume and/or capacity of production and/or field of use.
16 The use of quality controls by the supplying party, other than those required for trademark purposes.
17 The obligation to use a particular trademark, service name or trade name when using the technology supplied.
18 Obliging the acquiring party to provide equity capital or to allow the supplying party to participate in the management of the acquiring party as a condition to obtaining the technology.
19 Unlimited or unduly long duration of transfer of technology arrangements.
20 Limitations on the diffusion and/or further use of technology already imported.

Major differences still remain between developed and developing countries with regard to restrictive business practices. In relation to most of the practices on which general agreement has been reached, the developed countries propose the inclusion of a rule of reason formula. The disagreements on restrictive practices between developed and developing countries are profound. While developed countries insist that not all restrictive practices are unreasonable, developing countries believe that all restrictive practices are illegal by nature. Judging by current developments, it is not likely that in the near future an international law on restrictive practices that is acceptable to most countries in the world can be introduced.

SUMMARY AND CONCLUSION

This chapter has discussed the nature and major contents of restrictive business practices, antitrust laws of developed countries, technology transfer laws of developing countries and the approaches of the United Nations. On the whole, restrictive practices are obstructive to the development of technology transfer and international trade, and especially unfavorable to developing countries. Therefore, most developing countries regard restrictive business practices as irrational, thereby trying to exercise various controls on them. While developed countries also have antitrust laws to prohibit some of the restrictive business practices, they tolerate many of these practices to varying degrees.

In practice, restrictive business practices are a very complicated issue. While some of the restrictive business practices are definitely beyond the scope of patent protection and should be strictly prohibited in normal situations, many of them are derived from the valid concerns of suppliers and should be negotiated. A dogmatic approach toward restrictive practices can be harmful to the interests of both suppliers and recipients. Any attempt to utilize a favorable position in either transferring or receiving technology to deprive the other side of its legitimate right is counterproductive and harmful to the process of international technology transfer.

Part IV

Technology transfer via other major commercial channels

13 Other major commercial channels for technology transfer

INTRODUCTION

The international transfer of technology has regularly involved more than a bare transfer of industrial property rights or know-how. Frequently, these property rights and know-how constitute only part of a much larger arrangement which may comprise the provision of raw materials, skilled labor, engineering support, finance, management services and marketing assistance. Besides technology licensing, there are a number of commercial arrangements with which technology can be transferred across national boundaries. They include international franchising, international subcontracting, turnkey projects, counter-trade, R&D cooperation and technology swaps, joint production agreements, equity joint ventures, joint equity swaps and intrafirm transfer. The selection by the recipient of the proper form of commercial arrangement for the transfer of technology will depend on a variety of factors, ranging from trade and investment policies to the relative bargaining positions of the supplier and the recipient.

The level of economic development, for example, has significant impact on the form of transfer. If the recipient country is at the initial stage of industrial development with only primitive local manufacturing capacity, and the technology transferred is fairly complex, a turnkey agreement may be a very efficient channel of technology transfer. But when the recipient country has reached an intermediate stage of industrialization, or the technology to be acquired is fairly simple, an industrial property license may be most economical. Political and legal factors can also significantly affect the choice of transfer method by suppliers. Some of these are ownership regulations, the nature of economic planning, rules on the repatriation of capital and income, and the administrative mechanisms of the legal framework within which the transfer will take place. This chapter concentrates

13 Other major commercial channels for technology transfer

INTRODUCTION

The international transfer of technology has regularly involved more than a bare transfer of industrial property rights or know-how. Frequently, these property rights and know-how constitute only part of a much larger arrangement which may comprise the provision of raw materials, skilled labor, engineering support, finance, management services and marketing assistance. Besides technology licensing, there are a number of commercial arrangements with which technology can be transferred across national boundaries. They include: international franchising, international subcontracting, turnkey projects, counter-trade; R&D cooperations and technology swaps, joint production agreements, equity joint ventures, joint equity swaps and intrafirm transfer. The selection by the recipient of the proper form of commercial arrangement for the transfer of technology will depend on a variety of factors, ranging from trade and investment policies to the relative bargaining positions of the supplier and the recipient.

The level of economic development, for example, has significant impact on the form of transfer. If the recipient country is at the initial stage of industrial development with very primitive local manufacturing capacity, and the technology transferred is fairly complex, a turnkey agreement may be a very efficient channel of technology transfer. But when the recipient country has reached an intermediate stage of industrialization, or the technology to be acquired is fairly simple, an industrial property license may be most appropriate. Political and legal factors can also considerably affect the choice of transfer method by suppliers. Some of these are: ownership regulations, the nature of economic planning, rules on the repatriation of capital and income and the operative mechanisms of the legal framework within which the transfer will take place. This chapter compares various commercial arrangements from the position of the supplier of technology.

MAJOR CONTRACTUAL ARRANGEMENTS

While licensing and technical assistance are probably the most orthodox contractual arrangements for technology transfer, other contractual arrangements have in recent years become increasingly prominent in the transactions of international technology transfer, particularly with developing countries. The following is a brief introduction to the more important ones.

International franchising

Franchising is a variation of licensing in which a company (franchisor) licenses an entire business system as well as other property rights to an independent company or person (franchisee). The franchisee organizes its business under the franchisor's trade name and should follow the procedures and policies established by the franchisor. Under this form of business, therefore, the franchisor licenses the franchisee as a way of organizing and developing a business under its trade name in exchange for fees, running royalties and other compensation. Currently, one out of every three dollars Americans spend in retail establishments is paid to franchised businesses. The trend of development indicates that one out of every two consumer dollars will be spent on franchised services sometime in the next century (James and Weidenbaum 1993: 36).

Franchising has experienced explosive growth in the United States in diverse business fields, including car rentals, fast-food restaurants, hotels and motels, construction, soft drinks, and several services ranging from real estate brokerage to fitness centers. Franchised sales of goods and services at more than 520,000 locations across the country reached nearly $700 billion in 1990. There are a number of reasons behind the growing popularity of franchising. From the perspective of the franchisor, franchising constitutes an efficient way of rapid market penetration and product distribution without the large capital outlays associated with internal expansion. From the perspective of the franchisee, franchising provides a relatively safe way of owning a business, because it reduces the chance of failure as a result of the continuing training and support services offered by the franchisor. From the perspective of the consumer, franchised outlets provide a wide range of products and services at a consistent level of quality and at competitive prices (Sherman 1991: xii).

The collapse of Communism, a unified Europe, dynamic Asian economic development and the North American Free Trade Agreement (NAFTA) have combined to spur the growth of international franchising. Traditional US brand names such as Coca-Cola, Pepsi-Cola, Holiday Inn, McDonald's, Kentucky Fried Chicken, Dairy Queen and 7-Eleven (to name only a few) have become household names in many countries. In recent years, many US franchisors have expanded into the Asia-Pacific Rim. Japan, for example, has become one of the largest franchising countries in the world and is second only to North America with respect to the total number of franchisors and franchisees established (Chan and Justis 1990: 75). Although growing numbers of non-US companies have joined the international franchising business, US companies remain dominant in this form of business in both numbers and sales volume.

As with any other commercial arrangement, franchising offers both advantages and disadvantages to a company. In addition to the benefits of low costs for rapid expansion into foreign markets discussed above, the principal advantages include: a standardized way of marketing with a unique product or service image, franchisees with high motivation and commitment and relatively low political risks. Major disadvantages of franchising are similar to those of traditional licensing: limitations on the franchisor's profit, the possibility of creating competitors, inadequate control over the franchisee's operations and various governmental restrictions on the terms of franchise agreements.

International franchising is particularly enticing to a company when its product cannot be exported to a foreign target country, it is not interested in investing in that country as a producer, and its production process (or business system) can be transferred to an independent company in that country without much difficulty. Therefore, physical products whose manufacture entails significant capital investment and/or high levels of managerial or technical skills are usually not favorable candidates for franchising. The same logic is also applicable to service products that require sophisticated skills, such as accounting, advertising, banking, insurance and management consulting. This may be the main reason why international franchising is most popular in consumer service products that can be produced with relatively low levels of capital and skills (Root 1994: 134–135).

To be successful in international franchising, franchisors must do a great deal of homework before they go global. If franchisors are not careful and are ill-prepared for their international ventures, the current franchising boom may go bust. To avoid false starts, Steinberg (1993: 147–158) has offered the following advice to potential franchisors:

1 Examine the domestic operation to ensure that stability exists in the home market.
2 Refrain from taking action until the home operation is supporting itself from royalties.
3 Evaluate whether the basic concept has appeal in the foreign market.
4 Make a careful estimate of the real costs of doing business in foreign countries.
5 Ensure that the concept can work logistically and practically in the market.

International subcontracting

International subcontracting is a cross between licensing and investment, also known as outsourcing or contract manufacturing. In subcontracting, a company sources a product from an independent manufacturer in a foreign target country and subsequently markets that product in the target country or elsewhere. These contracts may be long-term, as part of a buyer–supplier relationship, or they may be temporary arrangements that terminate when the relevant contracts' activities come to an end. It was estimated at a very conservative level (which did not include intrafirm contract manufacturing) that worldwide contract manufacturing revenues were $22.3 billion in 1993, up approximately 30 per cent from the previous year (Marion 1994: 54).

Subcontracting encompasses agreements ranging from the purchase of components made overseas to be assembled at the home base of the company to the complete production of specific products by foreign manufacturers. To acquire a product manufactured to its specifications, the company is usually involved in the transfer of technology and technical assistance to the foreign manufacturer. These transfers can be formalized in a separate licensing/technical-assistance agreement between the two parties. Contracts concerning production by foreign manufacturers are to be differentiated from foreign direct investments, in which the parent company sets up and operates its own production facilities overseas (Kotabe 1989: 2).

Subcontracting has traditionally been important in a number of industries, including aerospace, automobiles and construction. As global competition puts growing pressures on firms to cut their costs and shorten product development time, subcontracting is becoming essential to many other industries, such as the electronic industry, because the development and production of new products involve such diverse technologies that a single company can hardly have competitive advantages in all of them. Heated competition has even driven rival firms into horizontal subcontracting agreements, each of which is capable of producing and marketing its products independently. Ford Motor Company, for example, obtains axles for its passenger cars and manual transmissions for its smaller trucks from Mazda (Spiegel 1993: 570–590).

International subcontracting may have the following advantages and disadvantages. The advantages are that it enables the company to enter into the target country without having to use significant financial and management resources; avoids local ownership problems; and allows the company to maintain control over marketing and after-sales service, which is hardly possible in the case of licensing and franchising. Major disadvantages are similar to those of licensing: it may be very difficult to find a suitable foreign manufacturer; even after an appropriate partner is located, significant technical assistance may be necessary to raise it up to the required quality and volume levels; the risk of creating a competitor is fairly high; and the company may develop an undesirable dependency on foreign partners as a source of key components.

Turnkey projects

In a typical turnkey project, a company designs, builds and installs capital equipment with the intention of turning over control and operation to the purchasers after an agreed period. For example, Fiat in the 1960s successfully delivered automobile-producing plants to the former Soviet Union and Poland, rather than selling automobiles. The result has been a second generation of automobiles based on Fiat technology, but Fiat had no control over distribution and sale once the original agreement had expired (Livingstone 1989: 88–89).

Turnkey projects are important in a variety of industries, typically involving the construction of large-scale capital works, such as the construction of water conservation systems, electric utility plant, food processing or food packaging plants, and pollution control, transportation, telecommunication and natural-resources exploration systems.

In the spring of 1991, for example, Pemex contracted Triton International Inc. (Houston) to carry out a turnkey project for oil exploration in southern Mexico's Gulf of Campeche.

While many turnkey projects are performed by companies from developed countries, an increasing number of companies from developing countries are participating in turnkey projects. Since 1980, for example, construction companies from Turkey have become a new force in the Middle Eastern and North African markets. After five or six years of subcontracting work, the Turkish firms moved into management consulting, turnkey project implementation and joint venture investment (Kaynak and Dalgic 1992: 60–75).

Five categories of turnkey agreement can be classified in accordance with the degree of involvement of the technology supplier in the planning and operation of an industrial plant (Blakeney 1989: 44):

1 A "turnkey contract" covers the construction and transfer of an industrial complex according to a design and incorporates technology to be selected and provided by a single contractor.
2 If a contractor is responsible for the major proportion of the establishment of an industrial plant, with other aspects to be arranged by the acquirer or subcontractors, a "semi-turnkey contract" will be negotiated.
3 Where the turnkey contractor undertakes to ensure that the industrial works will meet designated production targets under the operation of the purchaser, a "product-in-hand contract" will be signed.
4 If the purchaser chooses the technological process to be incorporated into a project, designed and constructed by the contractor, a "comprehensive contract" will be concluded.
5 If several contractors are to deliver different aspects of the project, supervised by the purchaser, "separate contracts" will be entered into.

One primary motivation for the providers to participate in turnkey projects resembles those of licensing and subcontracting agreements in that they allow companies to penetrate foreign markets that would otherwise be closed to export and foreign manufacturing entry. But as soon as the project is finished, the supplying company has hardly any control over the management of the facility or the marketing of the product. Furthermore, many turnkey contracts are signed with host governments. As such, they are very vulnerable to political risks of compulsory changes of key provisions, the unexpected calling of bank guarantees, or even contract abrogation. These risks should be fully assessed before a company enters into a turnkey project in a foreign country.

Counter-trade

In pursuing international business, companies are frequently faced with the challenges of having to accept all or part of their compensation in the form of goods, a business practice called counter-trade. In general, counter-trade may be defined as a contractual arrangement that involves a combination of trade, licensing, investment and finance. The parties to a trade transaction may engage in counter-trade upon the insistence of the importing country that it will not permit the import unless the original selling firm undertakes to export a corresponding amount of products from the country. In recent years, the term counter-trade has been expanded to include a wide range of additional business practices in which a seller provides other benefits to a purchasing country in order to obtain the sale contract of its goods, such as the use of local subcontractors, the undertaking of an investment in the country or the transferring of technology.

Counter-trade is conducted in a variety of forms (McVey 1985: 9–35). The principal forms of counter-trade include the following:

1 Barter or swap refers to the exchange of goods or services without the use of money, which is frequently introduced by the selling company as a means of "financing an otherwise unfinanceable transaction."
2 Counter-purchase is an agreement under which a selling company undertakes to buy products from the buying country which are unrelated to the items which it is selling; normally, the selling company does not have to fulfill its purchase obligation immediately after the delivery of the goods that it is selling.
3 Compensation or buyback is an agreement under which the selling company will sell industrial equipment, technology and/or an entire turnkey plant and agree to purchase a portion of the output manufactured from the use of the equipment or technology; a compensation transaction usually allows the selling company a significant period (ten years or more) to fulfill its obligation.
4 Offset can be defined as the agreement under which the selling company hires local subcontractors of the buying country or uses components or raw materials originating from that country; the offset agreement is often used for commercial aircraft sales,

defense-related contracts and others viewed as priority items by the buyer government.

For companies that wish to take full advantage of opportunities in developing and socialist countries, counter-trade may be a very useful and often the only way to penetrate these markets. Many of these countries have actively encouraged counter-trade. In spite of political and economic reforms in the former Soviet bloc and many developing countries, counter-trade still promises to be a significant way of consummating international transactions. And wherever necessary, technology transfer is routinely required as a part of the package. The offset transaction for F-16 sales between General Dynamics (GD) and Turkey is a good example. Faced with less costly competition from British Aerospace and the French aerospace companies, GD signed an offset arrangement with the Turkish government by offering to build a whole aircraft industry there and to invest in hotels and other projects as well. Under this offset transaction, GD bore the cost of "direct" offsets, i.e. exports, services and investment directly related to the sale. GD also paid for "indirect" offsets, which included general exports, services and investment not directly related to the F-16 sale (Schaffer 1989).

STRATEGIC ALLIANCES

Strategic alliances across national borders have become increasingly popular in some businesses. A strategic alliance is a long-term collaboration between two or more partners that combines their respective core competencies to develop global competitive advantage. A basic reason for such partnerships is to generate synergies not present when each of the partners works alone. A strategic alliance can involve cooperation on any business activity. The major motivations of partners are market access and resource sharing (including technology transfer).

R&D cooperations and technology swaps

Alliances based on the cooperation of companies engaging in R&D are different from those in which companies swap existing technologies, especially when R&D agreements pertain to the cooperation of companies either in the development of new technologies or in the improvement of existing technologies (Hladik 1988: 187). Parties to an R&D cooperative project may find it necessary to contribute technological skills, financing, management and equipment, etc. In contrast, technology swaps by and large resemble the licensing agreements discussed earlier. Instead of the transfer of technological know-how for money, technology swaps normally only refer to trades of technology for technology. Nevertheless, strategic alliances involving R&D exchanges regularly comprise both R&D and technology swapping.

Since the early 1970s, the number of cooperative R&D alliances has rapidly multiplied, as much as thirteen-fold from 1973 to 1988. Moreover, companies in the United States, Europe and Japan have tended to enter into more R&D cooperative ventures with foreign companies than with those in their home countries (James and Weidenbaum 1993: 72). The alliance between AT&T and NEC of Japan may be a typical example of the nature of many such R&D and technology-sharing partnerships. Under their agreement, AT&T provided some of its proprietary computer-aided design technology in exchange for advanced logic chip designs developed by NEC. Furthermore, both companies undertook to share the basic technologies on how to manufacture semiconductors as well as to collaborate in the development of more advanced semiconductor technologies.

Generally speaking, companies form technology-exchange alliances to share key resources and reduce costs that may be incurred in the process of technological development. Therefore, technology-developing alliances are very common in fast changing high-tech industries. Companies often participate in technology swaps when each company wants quick access to technology or know-how owned by the other side. Nevertheless, owing to the fact that neither the nature of the products nor the potential of future profits is predictable, there is a high risk for companies involved in such alliances. This risk may be derived from three sources: first, there is a real chance that the R&D or cross-licensing efforts may not lead to desired products; second, the market demand may be highly uncertain, especially for the introduction of new products; third, the future action taken by partners, especially if a partner is a competitor, may also be uncertain (Hladik 1988: 187).

Joint production agreements

Joint production agreements (or coproduction agreements) share notable similarities with both subcontracting and equity joint ventures, because these three arrangements regularly require the licensing of specialized elements (James and Weidenbaum 1993: 76). The main difference between joint production arrangements and subcontracting lies in the fact that coproducing partners are much more actively involved in the development of the product to be produced than is the case in typical subcontracting. Unlike equity joint ventures, the assets and management of each side are not incorporated into a joint business entity when the relevant partners undertake to engage themselves in joint production or distribution operations. While joint production arrangements may divert a significant share of resources from each side to the venture, their cooperative productions tend to be much less formally structured than an equity joint venture.

Coproduction is one of the most popular forms of interfirm cooperation between Western companies and their counterparts in developing countries or socialist countries, as many of these countries place various restrictions on foreign ownership of local companies. This is especially true in the aerospace industry, where coproduction is often a part of a counter-trade arrangement. On the one hand, local governments often require coproduction as a necessary pre-condition for the foreign aerospace industry to enter their markets. On the other hand, the development and production of a new model jet airline is so costly that even the largest companies in the industry may feel hard pressed for funds.

The transaction involving McDonnell-Douglas in China, for instance, is built around a coproduction program. The buyer and seller in this case collaborated to build an aircraft assembly system within Shanghai Aviation Industry Company. The coproduction program will transfer technology to the Chinese partner, not only to facilitate the assembly of planes, but to assist in the manufacture of some components for the plane. There is often a coproduction element in a large military offset, as is the case for GD in Turkey. In fact, coproduction is so significant to the concept of counter-trade that it is frequently the major goal of a foreign government's request for counter-trade (Schaffer 1989: 5). In the case of Boeing's new 777 commercial jet airliner, companies in six different countries were involved in sharing the financial burden and certain technologies for the development and production work.

Since joint production resembles joint venture and subcontracting, companies may base their decisions to form such a venture on similar considerations to those found in the other two arrangements. A joint production venture without equity can be very useful for those who may want to try out a partnership in order to assess its potential prospect of growing into a long-term alliance, with higher level commitment. If both partners are convinced that they have benefited from their cooperation and can work together for a long-term relationship, a higher-level alliance can be formed involving joint equity.

Equity joint venture

An international joint venture is started when a foreign company shares in the ownership of a company in a target country with local private or public concerns. As a common practice, a foreign company contributes capital and other resources together with a local company in a common endeavor. Depending on the equity share of the foreign company, joint ventures can be classified as majority, minority or fifty-fifty ventures. Under each one of these categories, the foreign company has less control over a joint venture than over a wholly owned subsidiary. A joint venture may be formed from scratch or via the foreign partner's acquisition of a partial equity ownership in an existing local company (Root 1994: 171).

Technology is often transferred to the joint venture through an assignment or through a licensing arrangement. The technology supplying partner may also be obliged to provide necessary technical and management assistance. In view of the perishable nature of technology, it is hardly surprising that partners pay considerable attention to devising technology transfer schemes to protect their respective competitive advantages, particularly when the effective joint venture will entail ongoing technical relationships between owners and the venture. If the joint venture requires inventive activity, the agreement will need to specify the respective financial, scientific and technological contributions of the parties to the joint venture, as well as the ownership rights of the partners in the industrial property rights and know-how which may be created (Harrigan 1986: 61–62).

As such, joint venture agreements inevitably comprise provisions on a number of the restrictions typical in licensing agreements. For instance, the markets in which the product of the venture can be exploited may be designated on a discriminatory basis, or third party competitors may be excluded from access to the products of the business or they may be required to pay discriminatory prices. Joint research venture agreements regularly contain restrictive provisions governing the confidentiality of the research activities of the venture, as well as specifying the markets in which relevant industrial property rights will be exploited (Blakeney 1989: 45).

One major reason motivating a foreign company to form a joint venture with a local company is the discouragement of wholly owned subsidiaries by the governments of some developing countries. Joint ventures may also be very beneficial to the foreign partner due to the local partner's contributions. These include capital, which reduces both the investment and the risk exposure of the foreign partner; management, production and marketing skills; expertise in the host country environment and business practices; close relationships with local suppliers, customers and government officials as well as excellent local prestige; and other resources. Among these benefits, the local partner's expertise in dealing with the local business environment is probably the most valuable. This has been the main reason for many foreign companies to select joint ventures in several Asian countries even though they can legitimately set up wholly owned subsidiaries.

On the downside, the foreign partner has limited control over a joint venture, especially when it has only a minority equity ownership. A joint venture can hamper a foreign partner's global and/or local strategy when the local partner's interests are at odds with its own. The formation of a joint venture usually requires complex and detailed contracts. It is also more difficult to protect intellectual property in a joint venture. Joint ventures are usually difficult to manage, due to differences in organization cultures and management styles. In contrast, a wholly owned subsidiary facilitates transactions between the parent and the venture, gives better protection of industrial property and promotes integration of the venture into the parent's worldwide system in terms of strategy, management and operations in other countries. With a wholly owned subsidiary, the foreign parent can also better implement its own strategy in the target country.

Joint equity swaps

Joint equity swaps, also known as minority investment alliances, resemble equity joint ventures in that they involve some form of co-ownership association between two or more companies. Under such associations, one firm obtains an equity position in another, usually holding less than majority ownership. Such equity alliances are typically designed to be long-term. Thus, companies refrain from creating joint equity swaps if the potential partners have not formulated a long-term strategy that will significantly benefit each firm. In cases where the two companies are traditional competitors, equity alliances have tended to be ineffective (Lewis 1990: 232–233).

Unlike an equity joint venture, a joint equity swap will not lead to the creation of a separate business entity. By forming an equity alliance, the parties hope to secure long-term commitment to cooperation. Unlike acquisition and merger, the exchange of equity between two or more firms allows partners to maintain a high level of separate identities. The main goal in exchanging equity rather than acquiring or merging companies is to promote closer cooperation between particular parts of each organization rather than to engage in an all-out cooperation (James and Weidenbaum 1993: 97–98).

IBM's purchase of a minority stake in Groupe Bull, the French state-owned computer maker, illustrates the uniqueness of such equity alliances. Through this agreement, IBM gained access to portable and notebook computer technologies while providing Bull with IBM's RISC-chip (reduced instruction set computing) technologies. Furthermore, IBM agreed to produce RISC products at manufacturing sites in France in return for Bull's commitment to make circuit boards for IBM.

INTRAFIRM TRANSFERS

Technology may be transferred to the target country within the context of transactions between an overseas parent company and a subsidiary established within that country. Relevant rights to intellectual property or know-how will be licensed to the subsidiary. Restrictions contained in these licenses will partly depend on the extent of ownership in the subsidiary held by the parent. The less ownership a parent has in a subsidiary, the less control it commands and the more restrictions it will impose on licensing. Usually, full ownership allows for the minimum restrictions. On the other hand, the subsidiary will be responsible for a designated activity within the corporate group. For example, it may be required to provide certain amounts of goods and services to be marketed by the parent. The extent and method of the transfer may also partly depend on the rules of the recipient country on foreign investment.

Since intrafirm transfers are usually less costly than interfirm transfers, a cost advantage can function as a bond. Several reasons for these lower costs exist (Keller 1990: 38–39): personnel in both countries share organizational and technological commonalities; time-consuming legal negotiations are not necessary; the risk of non-payment or a need for a performance bond is nearly non-existent; and concerns about a failure of the technology to work remain minimal. As such, a multinational corporation (MNC) often has a cost advantage in international technology transfer over two separate firms operating at arm's length. Philips of Holland, for example, has regularly employed intrafirm transfers among electronic components manufactured in Europe, Asia and the United States to keep costs low.

Meanwhile, there are several external pressures and company motivations for conducting R&D in a foreign subsidiary. Host governments have tended to both offer incentives and exert pressures on MNCs to conduct local R&D. Parent companies also often conduct R&D abroad due to subsidiary pressures. For instance, local staff might become frustrated if they remain just a manufacturing operation instead of a full-fledged member of the multinational family. For this reason, Sperry-Vickers acceded to the demands of its European subsidiaries for a fair share of R&D. Granting R&D to local operations may thus reduce discontent and boost morale in the subsidiary. In the transfer of technology among the large US MNCs, even though the initial corporate goal is to assist in the production of existing company products rather than in the development of new products overseas, the evolution of these subsidiaries has indicated the direction of new product development for local or even global markets (Cavusgil 1985: 226).

SUMMARY AND CONCLUSION

This chapter has briefly discussed and compared a few of the major commercial arrangements in which technology can be transferred internationally. The transfer options range from internal transfer to the firm's own subsidiaries abroad to external transfers via equity or contractual joint ventures, or to completely unrelated parties via collaboration agreements or other contractual arrangements. This, however, has not exhausted the full list of commercial arrangements in which technology can be transferred. International leasing can involve technology transfer, even though the financial arrangement is a dominant concern in leasing transactions. Product sales may sometimes necessitate the supply of technology to the recipient to assist the utilization of the goods. The acquisition of capital machinery may require technical help in its assembly, installation, operation and maintenance. Various forms of transfer of goods may entail the transfer of relevant industrial property rights to permit the utilization of those goods.

In terms of technology transfer, each form of arrangement has a number of advantages and disadvantages, with no single one being necessarily better than the others. In different environments, options available to international managers may vary a great deal. Thus, one has to compare these arrangements carefully before making an appropriate choice. Sometimes, a firm may place several arrangements under one package, as in a "spider's web joint venture," in which the firm may forge a variety of cooperative patterns to keep competitors at bay while improving its own position (Harrigan 1986: 3–4). Finally, once a choice is made, the company still has to handle the complexities of managing technology transfer. Proper precautions have to be taken to protect the transferred intellectual property. The following chapter will provide a good example of such complexities present in joint venture arrangement in a developing country.

14 International subcontracting and selected industries

INTRODUCTION

As already mentioned in Chapter 13, international subcontracting is one of the major modes of technology transfer. Also known as contract manufacturing and outsourcing, subcontracting "is a kind of half-way house between arm's length transactions on the open market and complete internationalization within the firm" (Dicken 1986: 189–190). A subcontracting relationship is formed "when a firm (the principal) places an order with another firm (the subcontractor) for the manufacture of parts, components, subassemblies or assemblies to be incorporated into a product which the principal will sell. Such orders may include the treatment, processing or finishing of materials or parts by the subcontractor at the principal's request" (UNIDO 1974). In parallel with the trend of globalization among domestic firms, international subcontracting has grown at an extremely fast pace since the mid-1960s.

In some major industries, such as electronics and automobiles, subcontracting has experienced such rapid development in the past decade that it has become dominant. In the electronics industry, for example, many big firms do not manufacture anything any more. Dell of Austin, Texas, and Zeos of Minneapolis are focused on marketing by simply purchasing circuitboards, disk drives and other modules – designed specifically for them – from outside manufacturers and assembling the computers in their own warehouses. Huge, vertically integrated producers like IBM and Apple are rapidly shifting toward the same model. The Apple PowerPC is fully subcontracted to Solectron's plant in Milpitas, California. In fact, most PC companies have subcontracted entire subassemblies to third parties (Tully 1994: 124). This chapter focuses on major forms of international subcontracting and technology transfer in international subcontracting, as well as subcontracting businesses in selected industries.

FORMS OF INTERNATIONAL SUBCONTRACTING

Subcontracting can be conceptually classified into various forms along different dimensions. Some researchers have examined a subcontracting relationship by studying the degree of transactional dependence either party has on the other. The degree of dependence of the buyer on suppliers may be influenced by the number of suppliers and the buyer's own in-house production capacity for the same products. The degree of supplier's dependence on the buyer varies a great deal according to the number of buyers (Jansson 1982).

Others have categorized subcontracting on the basis of the relative stability of a subcontracting relationship. They treat the relationship as a variable along a market versus a hierarchy governance continuum, in which the transactional nature of a subcontracting relationship varies from a short-term "free" market relationship at one end to a long-term stable partnership at the other end (Williamson 1986).

Still others have focused their attention on the contractual terms of a subcontracting relationship, which vary from fixed price contracts to cost-plus contracts. The differences in contractual terms directly influence the degree of risk shared between suppliers and buyers (Bamberg and Spremann 1987).

For the purpose of this chapter, UNIDO's classification (1974: 194), which divides subcontracting into three forms based on the principals' motivations, is adopted:

1 Specialty subcontracting, in which the contractor forges a permanent relationship with the subcontractor, who has specialized equipment or technologies.
2 Economic or cost-saving subcontracting, in which a contractor is mainly motivated by a subcontractor's considerably lower cost for certain production or processing.
3 Complementary subcontracting, in which a contractor looks for a subcontractor to make up for its insufficient production capacity due to sudden surges in orders.

But what constitutes the most important factors in "internationalized" subcontracting? Generally speaking, the international dimension of subcontracting is usually identified by one of the following two criteria: first, the principal and the subcontractor participating in the subcontracting relationship are from two different countries; or secondly, the subsidiary of a multinational corporation (MNC) enters into a subcontracting relationship with a local firm in the country of the subcontractor. As long as one of the two criteria is met, subcontracting can be deemed "international."

Michalet (1980: 47–49) has described four basic forms (cases) of international subcontracting: Case A – international subcontracting between two independent units located in countries at different levels of development; Case B – international subcontracting between the subsidiary of a multinational firm and a local firm or local firms; Case C – international subcontracting between the subsidiaries of several multinational firms; and Case D – international subcontracting between production units belonging to the same multinational group. These four forms of international subcontracting are related to different strategies pursued by the multinational firms involved.

Based on the nature of their transactions, Michalet (1980: 50–53) has further classified these four forms into two main categories: direct and indirect international subcontracting.

Direct international subcontracting covers Cases A and D shown in Figure 14.1. The common feature they share is the subcontractor's production which is completely exported to the principals. Irrespective of whether the subcontractors involved are purely local firms or a workshop subsidiary of a multinational firm, the products manufactured in the subcontracting country are designed to supply export markets rather than local markets. Through the agency of the principal, the product manufactured by the subcontractors will be channelled into markets in both developed and developing countries.

The way by which they are distributed mainly depends on the nature of the goods. If it is a luxury product with high demand elasticity, the share absorbed by the markets in developing countries tends to be small. It is not uncommon to find cases where the final product will later on be imported by the subcontractor's country of origin. Such a situation may happen even in cases where local subcontractors provide finished products such as clothing and television sets. This illustrates a prominent feature of subcontracting, i.e. the principals control the monopoly for marketing the finished product, which is protected by patents and licenses.

The subcontractor's dependence is deeper in Case D than in Case A. The standardization of products has to be more strictly adhered to. Ties between the different subcontracting units, which are subsidiaries of the same multinational group, have to be stronger, while links between the workshop subsidiaries and local industrial potential tend to be weaker. In contrast, independent subcontractors can often be the starting point for a chain of secondary subcontracting. Under such

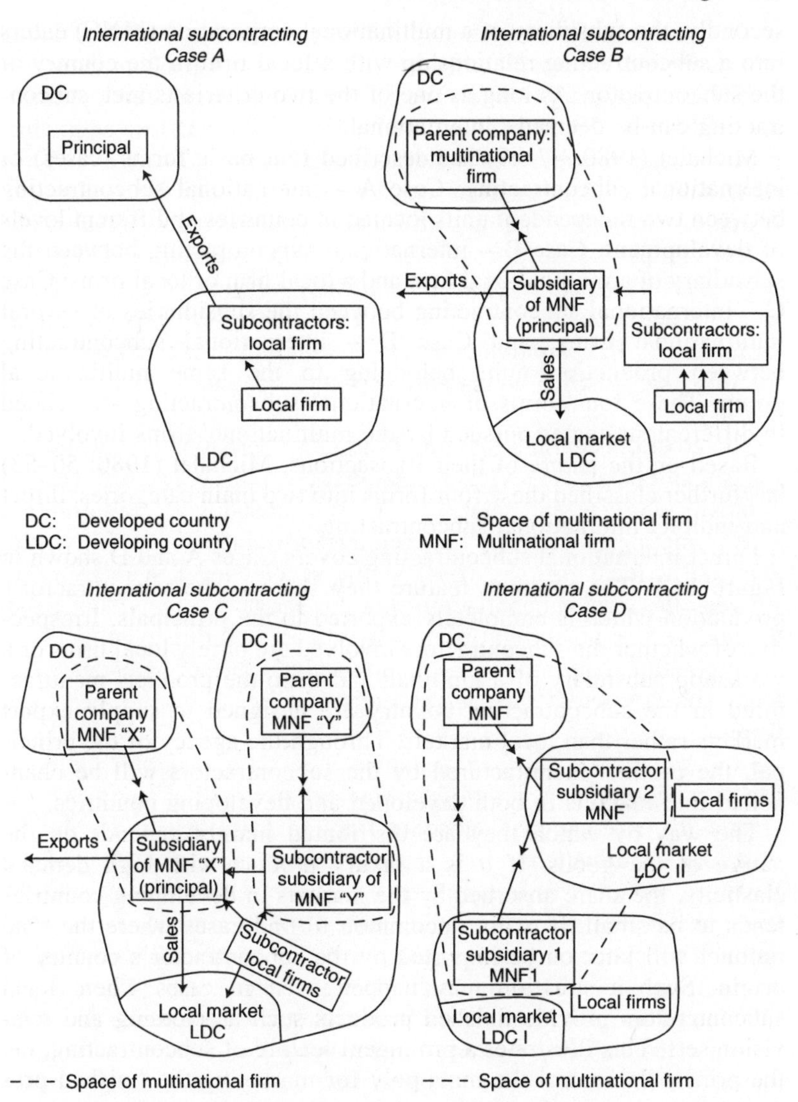

Figure 14.1 Four cases of international subcontracting
Source: Germidis 1980, pp. 51–52. © OECD, 1981. *International Subcontracting*. Reproduced by permission of the OECD Publications Service

circumstances, the geographical location of the subcontractors may be largely determined by the availability of skilled labor. Multinational subcontracting, for its part, will facilitate the standardization of production processes and labor.

The common feature which indirect international subcontracting (Cases B and C) shares with direct international subcontracting as embodied in Case D lies in the fact that it is organized by multinational firms. The workshop subsidiaries in Case D form a distinct contrast with the relay subsidiaries in Cases B and C, because the relay subsidiaries are mainly engaged in producing finished goods, which are exported and/or sold on the local markets. Usually, the parent company is responsible for worldwide marketing of the products manufactured or assembled by the relay subsidiaries. These products are finished and delivered in accordance with specifications laid down by the parent company. The range of products assigned to the relay subsidiaries varies with the development level of the host country.

The most outstanding feature of indirect international subcontracting is that it exists at two levels. The first level lies between the parent company and the subsidiary, while the second level represents a subcontracting relationship between the subsidiary of a multinational firm and a local firm or firms. In the second level, a relationship between the local subcontractors and the parent company does exist, though there is no straight principal–subcontractor connection. Such an international subcontracting relationship is further extended when local subcontractors themselves enlist the services of other local firms.

In short, the key distinction between direct and indirect international subcontracting lies in the principal's overall strategy. At one end is the international subcontracting in Case A, driven by the principal's motivations in its capacity as the direct producer of the finished product. The predominant concern is with the constraints of the domestic market. At the other end, Case D symbolizes an international subcontracting strategy that is based on the rationalization of production as viewed on a worldwide scale. Intermediate Cases B and C extend Case A to initiate the local market stage, as the principal becomes increasingly interested in breaking into foreign markets by manufacturing on the spot. It seems that Cases B and C have had a growing popularity among principals since the beginning of the 1980s.

MAJOR MODES OF TECHNOLOGY TRANSFER IN SUBCONTRACTING

In international subcontracting, technology can be transferred in a number of ways, depending on the nature of subcontracting. Technology transfer can take place in both directions to either contractors or subcontractors, although the flow to subcontractors has traditionally been dominant. Regarding the transfer from contractors to subcontractors, Wong (1991: 13–15) has identified five major modes (with the last one being bilateral exchange).

First, in an attempt to maintain and develop the subcontracting relationship, the MNC buyer may be obliged to make sincere endeavor and resource commitments to transfer certain technology of its own to its suppliers. This direct know-how transfer may include: advice on plant layout, equipment selection and operations planning; advice or training on quality management systems and other "good manufacturing practices"; on-site audits of plant operation and troubleshooting; loaning of equipment and machinery, either temporarily or permanently; training of supplier staff through formal courses and seminars or informal consultations or visitations.

Second, where the MNC buyer does not transfer know-how in the operation of the process technology of the supplier, its procedures to enforce its stringent quality/performance control system over the output supplied by the subcontractors may serve as useful feedback that may significantly enhance the technological learning of the latter. This learning facilitation effect may include: testing and diagnostic feedback on quality and other dimensions of performance of supplier's products; sourcing of technical experts to solve specific technical problems encountered by the supplier; and advance indications on future quality/performance/features requirements and targets. The more sweeping the MNC buyer's control, the more dynamic will be the learning process.

Third, the process of executing transactions through the subcontracting relationship itself tends unavoidably to involve the transfer of certain technological knowledge from the buyer to the seller. Commonly referred to as the "information disclosure," this indirect technology transfer or spillover comprises product design specification and performance requirements; early supplier involvement in the prototype development and value engineering stage; access to technical and market information on competitors' products; informal sharing of technical information and ideas among the technical staff of both companies; exposure to the MNC's system of managing and organizing manufacturing activities and observation of MNC corporate culture ("organizational and management technology"). Unlike the first mode, the technology transfer here is not consciously provided by the MNC buyer, and thereby hardly involves any active resource commitment from the buyer.

Fourth, when the supplier enters into the subcontracting relationship, it may be induced to make certain technological investments that it would otherwise not make without that relationship. Such an inducement effect can occur in several ways: the relationship may reduce the perceived risk of the investment decision due to implicit commitment from the MNC to purchase the new product or to increase the purchase of an improved product if the investment proves to be successful, to provide technical assistance if needed, etc.; the relationship may result in a relatively stable income that can be used to help finance the investment; the relationship may enable the supplier to access superior market demand information that leads to better investment decision.

Finally, as the technological level of the suppliers improves significantly, unilateral transfer may evolve into a bilateral exchange of technology, whereby the MNC buyers provide certain proprietary technology to selected suppliers in return for specialist technologies from the latter. This can be arranged in the form of cross-licensing of technologies, joint supplier–buyer R&D programs or joint venture agreements.

The relative importance of these modes depends on the specific nature of a subcontracting relationship. Moreover, Wong's categories are not mutually exclusive, with many subcontracting relationships having several modes at the same time. Although transfer efforts made by the MNC buyer have a significant impact on the extent of technology absorption by the supplier/ transferee, it is the commitment and ability of the latter to absorb the technologies that play a crucial role in the process of transfer.

It has often been argued that international subcontracting has not helped the technological improvement of suppliers. During the 1970s, such a view was prevalent in the developing countries. Based on a study of subcontracting relationships in the electronics industry, for example, a Korean scholar argues that the MNCs provide semi-finished products for mere assembly and sustain a very limited relationship with local manufacturers. Moreover, local firms that manufacture finished electronic products are usually required to import parts and components instead of using similar products made by foreign firms in Korea, because technology transfer contracts with foreign firms disallow the use of parts of other brands. While industrial growth has greatly contributed to the overall economic development of the country, the technology transfer pattern has not been conducive to a full-scale technology diffusion (Sang 1975: 118–122).

Others have argued to the contrary. Wong, for example, indicates that a supplier/transferee can seek to improve its technological sophistication along three generic technological ladders: "the process technology it uses to manufacture the products that it sells; the product technology that specifies the design of these products, and the quality assurance technology that ensures that the products comply with the quality/performance requirements of the buyer" (Wong 1991: 16). Beginning from low-level "know-what" of the product to be made and the process equipment/operation needed to manufacture it, the supplier can, with the help of product performance feedback from the buyer, move up to a higher level whereby he grasps a deeper understanding of the product design ("know-why") and/or an efficient operation of the process technology. A more advanced level is reached when he turns his specific product "know-why" into design "know-how" of other generically similar products and is capable of carrying out innovative adaptation of his process technology. At the most advanced level, the supplier can translate his knowledge of the particular product/process into an ability to innovate new products/processes entirely on his own.

For a more balanced view on technology transfer in international subcontracting, Watanabe (1980: 222) suggests that a distinction should be made between immediate effects, and long-term, sometimes indirect effects, when analyzing technological impacts of international subcontracting. Moreover, significant long-term impacts can be expected only when massive transactions occur. One important lesson he has drawn from the experience of East Asian economic developments seems to be that "a rapid quantitative expansion is an important cause, as well as effect, of qualitative improvement of not only industries but also of the economy as a whole."

Technology transfer can also take place in the direction from subcontractors to contractors. As global competition becomes increasingly intensive, contractors are more and more interested in the technologies developed by subcontractors. Companies operating in the global market can gain a sustainable competitive advantage by securing access to innovative technology developed overseas, thus locking competitors out of the technology base. This route to technical supremacy has already been aggressively followed in the electronics industry. In fact, subcontractors increasingly help contractors research and design distinct new products. Since subcontractors of components usually have previous experience with one part of a larger product, they can provide better designs (Fagan 1991: 22; and Tully 1994: 124).

As the pace, cost and complexity of technological developments become ever more challenging, globally oriented companies are changing their traditional way of managing their research and development activities. In an attempt to reduce costs and keep abreast of a widening range of relevant technologies, these companies are more and more dependent on external sources for R&D. In some extreme cases, companies have become completely reliant on external R&D. Japanese companies have taken the lead in contracting out R&D and already plan to increase their percentage reliance on external technology from 40 per cent to 60 per cent from 1993 to 1996. While most companies are still reluctant to outsource critical technologies, many of them have already begun to go outside their organizations for basic technologies (Houlder 1995: 14).

INTERNATIONAL SUBCONTRACTING AND SELECTED INDUSTRIES

Contract production thrives in the fast-changing world of the electronics industry. According to Technology Forecasters of California, in the worldwide electronics industry, subcontracting manufacturers collected $19.5 billion in revenues in 1992; by 1997, this figure could reach $40 billion. Component buyers (contractors) turn to subcontractors for several reasons: greater technical expertise, cost and quality controls, inadequate in-house capacity and growing pressure of customer demand. While dedicated contract manufacturers remain dominant in the industry, some electronic equipment manufacturers and distributors are also engaged in services. And computer production has become highly modular: workstations and PCs are clusters of separate sections, mainly the central circuitboard, disk drive, monitor and keyboard. Thus, subcontracting is very natural (Avery 1993: 57; and Tully 1994: 124).

In the past few years, many contract manufacturers have grown from low-tech board stuffers to high-level designers of advanced products. As they upgrade their technical skills and broaden their business scope to comprise purchasing, testing, software design, final assembly and distribution, subcontractors are becoming important members of product development teams at various levels of the electronics industry. Since technology changes are so rapid and the market is so volatile, these contract manufacturers are increasingly looked upon by buyers/contractors to rush products to market with the most up-to-date technology (Ristelhueber 1994: 48).

One outstanding example is the business software that has allowed

Company	1993 Revenue ($ millions)	Notable customers
SCI Systems	$1,679	IBM, Conner, Dell
Solectron	$ 836	IBM, HP, Sun
Avex Electronics	$ 510	Apple, AT&T, Dell
IBM	$ 450	Radius, Corollary, Dauphin
Bull	$ 350	Packard Bell, Compaq, HP
Jabil Circuit	$ 335	NEC, Quantum, Sun
Texas Instruments	$ 300	Sun, Compaq, IBM
Group Technologies	$ 244	Compaq, IBM, HP
Comptronix	$ 184	Apple, Ampex, GE
Kimball Electronics	$ 180	Lexmark, Kelsey-Hayes

Figure 14.2 The top 10 US contract manufacturers in the electronics industry
Source: Marion 1994: 54 (reprinted by permission of Elsevier Science Inc.)

IBM personal computers (PCs) to gain a strong market position, even though it was late in entering the market (Hibbert 1993: 67). The software used by IBM was not produced by IBM itself but was the product of the Lotus Development Corporation. Most of the components in the IBM PCs are produced by subcontractors whose special skills help bring PCs successfully to the market. IBM could not have produced its PCs in anywhere near the cost and time achieved via subcontracting had it tried to keep the production 100 per cent in-house. While Lotus provided applications software, Microsoft wrote the operating system on an INTEL microprocessor.

Meanwhile, IBM also provides contract manufacturing services. Four separate IBM units do contract manufacturing for both external clients and other parts of the company. With internal and external customers included, IBM's Canadian contract manufacturing unit, Celestica Inc., itself had a revenue of more than $1 billion in 1993. In fact, IBM's contract manufacturing relationships are so complicated that even its top managers have trouble figuring out actual revenue. Some estimate IBM's overall external contract manufacturing revenue at more than $200 million while others believe that the total may be as high as $450 million (Marion 1994: 54).

Subcontracting has long been the norm in the automobile industry's operations (Helper 1991: 15–28). The recent recovery of Detroit has provided a major boost to contract manufacturing. Like the computer industry, automobile manufacture is becoming highly modular: suspension specialist A. O. Smith, window maker Donnelly and seat manufacturer Johnson Controls ship their products straight to the assembly lines of several car companies. Chrysler now sources 70 per cent of the parts in each car from suppliers while making only the engine, transmission and metal skin for its LH series – the Chrysler Concorde, Dodge Intrepid and Eagle Vision. The trend toward subcontracting is so powerful that even the most vertically integrated General Motors, which runs a $25 billion parts business and manufactures 70 per cent of its components in-house, is now closing plants and outsourcing production (Tully 1994: 127).

The Ford Escort is a good example of the extent of international subcontracting in the automobile industry (Hibbert 1993: 73). In the mid-1980s, the Ford Escort was assembled in Great Britain and the Federal Republic of Germany, but its components were sourced from a global network of suppliers, with some being part of Ford Motors and others independent subcontractors. Figure 14.3 represents the network around Ford Escort. The company made a long-term strategic decision to depend on international subcontractors for these components and the technologies that were used to manufacture them. Behind this strategic decision lies the company's commitment to close buyer–supplier collaborations. The company was also prepared to integrate product development and total quality management (TQM) with the network of suppliers.

Subcontracting is a common practice in the aerospace industry, where the complexities and costs of production necessarily require the contracting of partial operations to specialized companies. Boeing Co., for example, has subcontracted various parts of its planes for several reasons. By subcontracting peripheral production to other companies, it can specialize in the design or production of core parts or technologies of aircraft. By subcontracting internationally, it also has a better chance of gaining foreign markets or significantly reducing its production costs. As new breeds of Boeing aircraft involve ever more complicated technologies and production processes, international subcontracting has become indispensable.

Boeing's production of the 777 serves as an example (AW&ST 1991: 41). Boeing selected three main subcontractors to supply major portions of the 777 aircraft under lengthy contracts that cover products incorporating advances in technology. The agreements, valued at almost $1 billion, involved companies in three countries. The

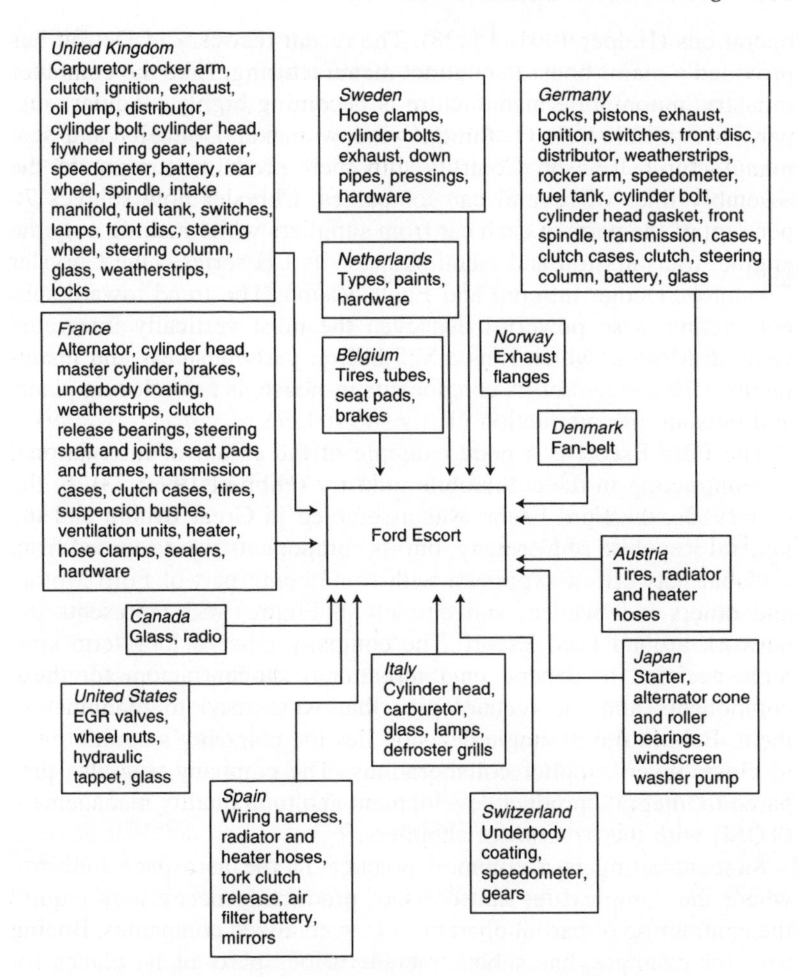

Figure 14.3 International subcontracting in the automobile industry
Source: Hibbert 1993: 74

agreements comprised 500 shipsets and were expected to last more than a decade, as the new industry trend encouraged long-term relationships between prime contractors and suppliers.

Under the contracts, Menasco Aerospace Ltd, of Oakville, Ontario, Canada, would supply the main landing gear – the largest ever designed for a Boeing aircraft – as well as the nose gear. Menasco would subcontract some assembly and manufacturing to Menasco Aerosystems of Fort Worth, Texas, and Messier-Bugatti of Velizy,

France. Rockwell International's North American Aircraft Operations, Tulsa (Oklahoma) Facility, would supply graphite composite floor beam assemblies, which could reduce weight by 28 per cent compared with conventional materials. Grumman Aerospace Corp. would provide the graphite composite spoilers, which would be produced by its Aerostructures Division at Milledegeville, Georgia.

SUMMARY AND CONCLUSION

This chapter began with a discussion on major forms of international subcontracting, with a focus on Michalet's four cases. This was followed by an analysis of major modes of technology transfer in international subcontracting. Five major modes of transfer have been explored. While the transfer from contractors to subcontractors remains dominant, the reverse transfer is gaining momentum. At the end of the chapter, international subcontracting in selected industries (electronics, automobiles and aerospace) were examined. Prominent cases in each of the three industries were discussed. Depending on the specific environment, both principal companies (contractors) and suppliers (subcontractors) can gain significant competitive advantages by participating in international subcontracting. It is generally expected that international subcontracting will continue to grow in the years to come.

As mentioned in the last chapter, international subcontracting also has some disadvantages. In order for contractors to take advantage of international subcontracting, Hibbert (1993: 75) has identified several basic criteria for selecting subcontractors, including, among others: "flexibility and gearing toward on-time delivery"; "ability to meet quality standards and implement TQM"; "financial soundness and effective management"; "ability to integrate with the manufacturer's business and to develop new expertise as required in design and so on"; "recognition of the concept of value for money in long-term or contract supply arrangements"; at a higher level, ability to "offer design and manufacturing expertise in specialized fields superior to that of their customers"; and ability to handle "sudden fluctuations in demand and other non-recurring events."

15 Transferring technology to China via joint ventures

INTRODUCTION

Since China opened up to foreign investment in 1979, it has attracted approximately US$26.7 billion direct foreign investment, with approved enterprises reaching 37,000 in 1991. Equity joint ventures accounted for about one-third of the country's total direct foreign investment, and were favored by foreign investors during the period 1979–88 (*Beijing Review*, March 23–29, 1992: 42). In spite of various difficulties, many equity joint ventures have obviously thrived in China. With a realistic plan and good preparation, one can reasonably expect success in creating and managing a joint venture in China.

The primary Chinese motivations for entering into industrial joint ventures are the acquisition of advanced technology, foreign exchange and management expertise. While there are many vehicles for the transfer of technology, such as licensing, coproduction and subcontracting, the equity joint venture is popular because it allows the foreign partner to realize its major objective, i.e. to participate in the Chinese market while maintaining some control over business activities. This chapter examines several key aspects of creating and managing joint ventures in China, with a focus on issues related to technology transfer.

FINDING A GOOD PARTNER AT THE RIGHT LOCATION

As in any long-term relationship, the first step toward success is to find a good partner. If a foreign investor prefers an equity joint venture to other forms of investment, he should first of all lay down the criteria and requirements to be met by his potential

Chinese partner. This seems to be fairly easy on the surface, as Article 1 of the Law of the People's Republic of China on Chinese–Foreign Joint Ventures merely requires the Chinese partner to be an entity enjoying the status of a legal person. But to find a suitable partner is not an easy job.

To begin with, one needs good connections (*guanxi*) to be introduced to the right partner. *Guanxi* is one of the major dynamics of Chinese society. The term refers to special relationships two people have with each other. It can be best translated as friendship with implications of a continual exchange of favors. On the whole, *guanxi* binds people through the exchange of favors rather than through expressions of sympathy and friendship. But *guanxi* requires time and patience to cultivate. To establish a good connection network in China is probably the most worthwhile yet the most daunting task to fulfill. Without *guanxi*, one often, in a Chinese phrase, "gets half the result with twice the effort."

But *guanxi* is not the only thing necessary to the success of a joint venture in China. One should have knowledgeable ideas about the specific conditions of the potential Chinese partner. The following questions are fundamental:

1 What is the financial situation of the potential Chinese partner?
2 What ownership does the Chinese partner have: public, collective or private?
3 Are leading members of the Chinese partner professionally competent and devoted?
4 Does the Chinese partner receive strong backing from the relevant governmental institution?
5 How much power does the partner have in obtaining essential supplies and raw materials?
6 What kind of access does the Chinese side have to China's domestic market?
7 Are competing ventures planned elsewhere or does the Chinese side have the political and economic clout to keep a dominant position in its field?
8 Can the Chinese side successfully assimilate the transferred foreign technology?

There are a number of common pitfalls in locating a suitable Chinese partner. One is that some potential Chinese partners are in serious financial trouble. Joining forces with a foreign partner may serve to rejuvenate a dying organization; this is particularly the case with some state-owned enterprises. Some Chinese call this strategy "fishing": to hook unsuspecting foreign investors by hiding the dismal financial and management problems of the enterprise. Once the initial investment is made, the foreign investor will feel obliged to spend additional money to ensure that the ailing enterprise will be profitable. William Mallet of Tianjin, China, advises that, "Foreign investors shouldn't try to resurrect worthless organizations. They should link with ones that are strong financially, because that will provide a good foundation" (Goldenberg 1988: 183–84).

Another pitfall lies in differences amongst priorities. Generally speaking, foreign partners are interested in market access, cheap labor and lax rules on pollution control while the Chinese side is interested in capital and technology as well as promoting its exports. When these various priorities are at odds, coordination between the joint venture partners is very poor. This is the so-called situation of "sleeping in the same bed and having different dreams," as a Chinese phrase goes. Beijing Jeep typified one of these clashes over priorities between foreign and Chinese partners. With a strong interest in absorbing technology, the Chinese felt that American Motors Corporation (AMC) had reneged on the terms in their contract calling for joint design and production of a new Jeep. AMC suggested postponement of further work when they found that the exhaust system, noise controls and speed failed to meet international standards (Mann 1989).

Having said all this, there are many success stories and a suitable Chinese partner can be very helpful in a number of ways. They can significantly cut the costs of production, help the overseas partner to break into the enormous Chinese market, and even contribute to technology improvement of the parent company. In many areas, a good Chinese partner is almost indispensable for the success of a business operation in the country. The operations of the McDonnell Douglas cooperative venture and the Johnson & Johnson venture in Shanghai are two notable examples of success. In both cases, Chinese partners played a key role in promoting the products in the Chinese market.

The right location is also crucial. A recent survey reveals that Shanghai and Shenzhen are home to China's most profitable American joint ventures, with their respective returns on investment (ROIs) at 16.2 per cent and 13.6 per cent. In contrast, the inland cities seem to have been the most disappointing locations for American investors. Average ROI in these locations is only 10.1 per cent, lower than the average coastal or Special Economic Zone (SEZ) rate. Moreover, nearly half of the executives reporting on inland investment projects responded that their joint ventures have not met expectations due to less-developed infrastructures, poor transportation networks and uncertain raw materials supplies (Stelzer *et al.* 1991: 54–55).

Nevertheless, being in an inland town or province with a committed municipal or provincial leadership can also have some unusual benefits. There are already many foreign investors in Shenzhen, Shanghai and other coastal cities, but relatively few ventures in many inland cities. The costs and expenses of operating a manufacturing facility can be considerably lower in an inland province than in the coastal areas. Joint ventures located in these regions can get the personal attention and support of the provincial governor or mayor. Closer personal interactions with top local decision makers can be easier and more frequent. These interactions have become increasingly important during the past decade as the Chinese economy has been considerably decentralized, thus shifting substantial power to local governments.

There are many cases in which foreign ventures in some small inland cities have managed to overcome mounting obstacles due to direct intervention from local governments, which normally cannot be expected from the coastal cities' governments. Many local officials are eager to attract foreign investment. In order to offset their geographic disadvantages when competing for foreign investment with coastal areas, some of them tend to adopt aggressive policies. Many go out of their way to be personally involved in supporting joint ventures and the commitment from local government is indispensable to the success of a joint venture.

PREPARING FOR NEGOTIATION

One of the first things a potential investor needs to do is to familiarize itself with the legal framework for technology transfer in China (Goosen 1987). Currently, the Regulations on the Administration of Technology Import Contracts (RATIC), promulgated by the State Council on May 24, 1985, are the most comprehensive. To facilitate and streamline China's technology import, MOFTEC (Ministry of Foreign Trade and Economic Cooperation, formerly MOFERT, Ministry of Foreign Economic Relations and Trade) promulgated on January 20, 1988 a set of rules, the "Detailed Rules and Regulations for the Implementation of the Regulations on Administration of Technology Import Contracts" (Detailed Rules).

The Law of the People's Republic of China on Joint Ventures Using Chinese and Foreign Investments (The Law) as adopted at the Second Session of the Fifth National People's Congress, July 1, 1979 and the Regulations Governing the Implementation of the Law of the People's Republic of China on Joint Ventures Using Chinese and Foreign Investments (The Regulations), promulgated by the State Council on September 20, 1983, are the two most important laws for those who want to transfer their technology via joint ventures.

Relevant tax laws should draw special attention from potential investors. Several major changes have been made by the government, and more changes are coming. According to the Unified Income Tax Law of 1991, joint ventures must pay a flat 30 per cent income tax on net income to the central government and another 3 per cent to the local government, totaling 33 per cent. When the foreign partner of a joint venture remits its share of profit home after tax payment, it no longer has to pay a further remittance tax of 10 per cent. For joint ventures whose exports exceed 70 per cent of their total output or are located in the special economic zones or in the technology development zones in the open coastal cities, income tax is cut by half to 15 per cent. For joint ventures with a contract for more than ten years, there will be no tax for the first two profit-making years, and there will be a 50 per cent reduction of income tax for the following three years. Joint ventures which reinvest their profits in China over five consecutive years will have a refund of up to 40 per cent of the tax payment for reinvestment ("United Income Law" and Peck 1991: 12–15).

Like other investments in China, joint ventures must be approved by the relevant governmental institution, depending upon the scale of investment. The most recent rule allows local governments at provincial and municipal levels to authorize joint ventures with less than $30 million in investment. The first step is for the Chinese partner to present a proposal for the joint venture to the foreign investment commission of the local government or to MOFTEC if the investment involves more than $30 million. The following items should be provided by the foreign side: the name, country of origin and legal address of the foreign firm as well as its scale of operations and business standing; the proportions of the capital investments by the foreign partner (normally no less than 25 per cent) and the types of investment; the major products of the joint venture, and the proportions of international sale ("The Law" and Ho 1990: 22).

This is followed by a feasibility study, which normally includes: economic significance of the investment; market demand and the scale of the proposed product; necessary resources, fuel and public utilities; proposed project location and project design scheme;

environmental protection; organization of production; employment and training of workers and staff; implementation plan; investment estimate and manner of raising funds; and estimated product cost and economic results. One common complaint from foreign partners is that in order to impress them, some Chinese partners tend to verge on over-optimism, changing the study of feasibility into a study of the "imagination." Although neither party is obligated to sign a contract that conforms exactly to the feasibility study, the study does have normative power in two areas. First, in order to deviate from the feasibility study, the party which wants to deviate must obtain the consent of the other party and the approval of the government. The burden of proof lies heavily on the party which wants the change. Second, throughout the duration of the joint venture, the feasibility study may be referred to in situations where the contract language is ambiguous (Fischer 1993: 12).

Upon approval of the proposal, both sides can begin to draft the contract for the joint venture, which is the key document of the venture and must be in both Chinese and the language of the foreign partner. The contract should also include the duration of the joint venture, the share of roles and power among the executives, the means to resolve disputes and the way to dissolve the joint venture ("The Regulations" 1983).

Negotiations on the contract can be long-winded. With a tendency to negotiate in a style described by one veteran China trader as "a blend of the Byzantine and evangelical," Chinese negotiators often frustrate Western businesspeople unused to such tactics. Many Western negotiators leave the table feeling pessimistic about their future partners. Several tactics are systematically used, such as controlling location and schedule, utilizing weaknesses, using shame tactics, pitting competitors against each other, feigning anger, rehashing old issues and manipulating expectations. Often negotiation continues after the signing of the contract. Therefore, a foreign partner should choose the right negotiators, prepare for time-consuming rounds and develop a sophisticated strategy before starting negotiation (Chen 1993: 12–16).

DEALING WITH SPECIFIC PROBLEMS IN NEGOTIATION

One major problem at this stage is that foreign suppliers often find themselves in a dilemma whereby the Chinese demand technology that is "state-of-the-art." Chinese negotiators routinely request the "most advanced" technology from foreign suppliers during initial negotiations, even though they may lack sufficient foreign exchange and have an inadequate infrastructure to utilize complex technology. A McDonnell Douglas Helicopter executive mentioned to the author that after he explained how easily new model MD helicopters can be maintained, his Chinese friend became annoyed and privately advised him to describe the maintenance as a complicated process in order to show the "quality" of technology.

Not only must the technology be the most advanced, but it must also be the most price competitive. Price negotiations on the transferred technology are one of the most daunting challenges to foreign negotiators. The Chinese frequently complain that the valuation of foreign technology is placed too high while foreign managers argue that their advanced technology is very expensive to develop (De Bruijn and Jia 1993: 19). Whoever can convince the Chinese to accept appropriate technology and package the transferred technology in the most competitive way holds the key to a successful negotiation.

Another persistent problem is how to decide on the contribution made to the venture by each partner. Foreign partners have often found that the Chinese insist that they pay a large sum of foreign cash to the venture. On the other hand, the Chinese side prefers to contribute non-cash items, such as land use, existing buildings and construction materials, all of which are easy for the Chinese to overvalue due to the difficulty in assessing prices accurately. One manager that the author interviewed in 1994 complained that the value of an out-of-date computer was estimated at close to its original purchase price, even though the foreign partner had to buy one to replace it. In order to avoid such complications, foreign companies should take advantage of assessments by professional consulting companies.

When investing their technology in a joint venture, the suppliers may have some additional complications. Although this last option is an important means of technology transfer worldwide, capitalization of technology is often discouraged by Chinese authorities due to their fear of underfinanced joint ventures in which foreign partners mainly contribute intangible assets. Chinese recipients are often reluctant to accept technology as a significant contribution of capital because Chinese firms are subject to various "internal policies" of the government, which often restrict the capitalization of technology to a 15 to 20 per cent maximum. Although such policies are not publicized, projects that capitalize the majority of the technology tend to slow down the process of approval or even lead to its eventual disqualification. Worries about improper capitalization of technology have led the Chinese to be more inclined toward technology licensing when the supplier is a joint venture partner.

Two more issues have proved to be especially cumbersome. The RATIC and its "Detailed Rules" impose strict restrictions on tie-in arrangements. Unless special approval is received, the recipient cannot be required to purchase "unnecessary technology, technical services, raw materials, equipment or products," nor may its freedom to purchase "raw materials, spare parts or equipment" from sources other than the supplier be restricted. In order to reduce hidden costs for "unnecessary" ancillary services and equipment, stipulations against such arrangements are often included in technology transfer control laws.

Such regulations present a serious problem to potential suppliers because such tie-in arrangements provide a major incentive for foreign companies to transfer their technology. It is also often impossible for a supplier to fulfill contractual guarantees requiring that the technology "achieve the objectives stipulated in the contract," unless the recipient purchases the necessary technology and makes use of appropriate equipment and raw materials in the application. Therefore, the tie-in arrangement is necessary to protect the legitimate interests of the supplier. Without such an arrangement, contract provisions would have to add that if inputs are not properly sourced according to the supplier's specifications, the supplier shall not be liable for problems caused by substandard inputs.

The RATIC and other related Chinese laws also prohibit suppliers from placing restrictions on quality, price or region of sale and export. Similar provisions prohibiting scope and volume of production constraints, as well as price fixing are expressed in the technology transfer regulations of many developing countries. Suppliers may indeed have some legitimate concerns because one of the principal purposes of these clauses is to protect the sales based on which royalties are calculated. In reference to exports, a restriction is clearly "unreasonable" if it prohibits the recipient from exporting to a market where the foreign supplier has no presence. Suppliers again do have legitimate cause for protecting their existing markets. Fortunately, Chinese authorities often accept restrictions that prevent direct competition with the supplier's own products in their home market (Cohen and Pierce 1987: 46).

The issue of intellectual property can be a vexing problem unless proper measures are negotiated to secure protection. The Memorandum of Understanding (1992) and the Agreement on Intellectual Property Protection (1994) have provided a significantly improved environment for the protection of investors' intellectual property (see Chapter 8). Nonetheless, pitfalls still exist. For example, RATIC allows a Chinese recipient to simultaneously use other foreign suppliers' technology. Various provisions are found in technology transfer laws of other developing countries which challenge the validity of clauses restricting the recipient's use of competing technology. This is perceived to hinder the recipient from making a reasonable selection of appropriate technology. When the Chinese recipient cooperates with a direct competitor of the foreign supplier it is more difficult to maintain trade secrets. Moreover, the turnover rate of Chinese employees in joint ventures, especially at the managerial level, is fairly high, due to the shortage of qualified managerial staff. In the light of these potential complexities, careful drafting is integral to giving the supplier proper protection.

Once the contract is approved, the Chinese side will need to petition the local office of the Administration of Industry and Commerce for a license for the venture in order to begin operation. The joint venture is not officially recognized until it receives this license. Upon granting of this license, the joint venture will need to contact the tax agency, get registered in the customs office, and open accounts in foreign and local currencies. After the completion of all these activities, it can begin its operations and recruit workers.

FOREIGN EXCHANGE BALANCE AND LOCALIZATION

The balance of foreign exchange is crucial to the success of a joint venture. According to the Joint Venture Law, foreign exchange inflows and outflows should at least balance over the life of the joint venture. Chinese authorities are very sensitive to the projected foreign exchange balance of a joint venture. According to a recent survey, high-tech/export producers tend to show the highest return on investment (15.4 per cent), with the highest number (58 per cent) of respondents indicating that their ventures have generated higher-than-expected returns. Those that produce import substitutes have the lowest return (8.7 per cent) and are the least satisfied, with 45 per cent of those respondents indicating that the results have been below expectations, perhaps reflecting the higher priority the Chinese government has given to export-oriented businesses (Stelzer *et al.* 1991: 56).

Basic principles on foreign exchange are as follows. A joint venture should normally show in its feasibility study a positive foreign exchange balance by the end of the term of the contract. Foreign exchange can be used for necessary imports, but a strong preference is given for local substitutes. Foreign employees in the joint venture can be paid in hard currency. The payment of dividends to the foreign partner can be in hard currency. These are some of the general principles which can be applied with varying degrees of flexibility, depending on China's overall reserve of foreign exchange. As a rule, pro forma financial statements that project early foreign exchange surpluses tend to facilitate negotiations and operations. A joint venture must submit periodic reports of its foreign exchange balance to the State General Administration of Exchange Control (SGAEC) ("The Provisional Regulations" 1980). The best strategy to keep a good balance of foreign exchange is to insure an international market for at least 50 per cent of the products.

The curtailment of foreign exchange could cause tremendous problems for joint ventures depending on foreign exchange to pay for imported parts. Beijing Jeep, for example, failed to obtain foreign exchange from foreign residents in China as it had originally planned, because they either had other priorities for their funds, or travelled in imported company cars or used the reasonably priced taxis. Moreover, a Beijing Jeep costs nearly twice as much as its US cousin. The company was virtually stranded until the Chinese government, worrying about the unfavorable impact of failure for a well-known venture on foreign direct investment in the country, opened up China's market to the Jeep venture (Goldenberg 1988: 96–97).

Special provisions have been made for customs duties on import and export commodities of joint ventures. Equipment, transportation equipment, office facilities and other supplies can be imported duty-free as part of the original foreign investment specified in the contract. Additional purchases, approved by the authorities concerned, are also duty-free. A joint venture can either make these purchases from the Chinese market or from abroad. For domestic purchases, it pays in Chinese currency at prices which are the same as those paid by the Chinese state-owned enterprises. For international purchases, the prices charged to the joint venture should conform to comparable international market prices. Obviously the joint venture that can take advantage of Chinese supplies will enjoy lower material costs.

But sourcing for production in China is no easy task. The quality standards that foreign partners require usually surpass those available from Chinese suppliers. When a Chinese supplier's quality is acceptable, its products are in high market demand, which causes problems in securing a steady supply and requires payment in foreign currency.

Otis, for example, had tremendous difficulty in locating quality local sources. Localization proved to be the most difficult task for the company (Hendryx 1986: 57–66). It was anticipated that the lack of identical locally produced, equivalent materials or components would require that either substitutes be selected – which would necessitate re-engineering if the materials or components differed significantly from the original – or parts be imported.

Apparently, the success of localization is a key factor to the success of joint ventures. To achieve rapid localization, the Beijing Jeep Corporation made heavy investments to import the American engine production line, the British printing system, the Japanese mold and crucial tools and the DISA2120 automatic production line from Denmark. Meanwhile, Beijing Jeep Corporation set up a Cherokee components localization commission with other Chinese firms to speed up this localization. Shanghai Volkswagen Co. Ltd took advantage of Shanghai's regional industrial advantages to successfully expedite Santana and Audi localization production. Empirical evidence indicates that localization is also a crucial step in the transfer of technology (De Bruijn and Jia 1993: 20, 22).

CHALLENGES IN HUMAN RESOURCES MANAGEMENT

Personnel management is also vital to the success of a joint venture, particularly for the joint ventures involving technology transfer. Many of these companies need intensive training to ensure a smooth process of transfer and the production of good quality products. One major incentive for foreign companies to invest in China is its relatively low labor cost. If they can manage the workforce well, they can take advantage of low costs and have a better chance of success. To attract foreign investment, the Chinese government has authorized a series of prerogatives for joint ventures in hiring, supervising and firing employees and in determining the amount of compensation ("The Regulations on Labor Management" 1980).

In the words of a general manager of a diagnostic reagent manufacturing joint venture in China, three factors are key to the success of any venture in China: "people, people and people" (Epser 1991: 24–30). According to his experiences, several rules should be kept in mind when hiring Chinese staff:

1 Avoid taking too many employees from a single source, as this can heighten the risk of hiring a lot of people with similar bad habits who may reinforce each other.

2 Practice patience and flexibility in looking for high-quality personnel.
3 Resist pressures by Chinese authorities to overhire.
4 Find a confidant among the local management staff who has experience in dealing with the bureaucracy and is trustworthy.

A joint venture is required by law to inform the local supervisory agency and the local labor department of the types and numbers of positions open. Any vacancies should be announced publicly and an open test should be held for the candidates to compete on an equal basis. Employees are subject to a probationary period. Employees' wages are usually determined on the basis of the qualification, workload and contributions to the joint venture. Subject to advance notice, a joint venture may lay off employees either for their poor performance or because of serious losses on the part of the venture. On the other hand, employees may also resign with an advance written notice. In fact, the high turn-over rate of highly qualified and well-trained employees has been one of the most challenging problems for joint ventures.

One common problem that has been identified by many Western managers is the difficulty in getting their Chinese managers actively involved in the decision making process. Japanese managers, on the other hand, have tended to complain about their failure in practicing the consensus decision process. The traditional dynamics of Japanese-style management, where production efficiency, quality control and marketing expansion lean heavily on the loyalty of employees, are not easily transferred to the Chinese (Ireland 1991: 14–17).

Many Western companies which transfer technology via joint venture have also encountered the problem of maintaining confidentiality. Although Chinese law provides that the licensee (the joint venture) must adhere to the confidentiality provisions of the contract, confidentiality remains a worrisome problem, as employees may attempt to appropriate proprietary technology after they leave the company. Therefore, the contract should provide that disclosure of confidential information to employees shall only be on a need-to-know basis. Each employee to whom such information is disclosed should sign an individual confidentiality agreement with the venture (Cohen and Pierce 1987: 47).

A good relationship with the trade union proves to be very important in many cases. Like their Western counterparts, Chinese trade unions are supposed to protect the welfare of the employees. The unions, for example, have the right to set up a reserve fund for employees' self-improvement in job-related skills, as well as for recreational activities. A joint venture is required by law to pay monthly contributions to the union reserve fund, equivalent to 2 per cent of the total wages of the employees who are members of the union. The unions have the right to attend meetings of the Board of Directors when issues concerning the workers are discussed and to appeal to the board on behalf of workers with regard to disciplinary action or dismissal. But in a somewhat different vein than Western unions, the unions in China have a distinct responsibility to urge employees to comply with the rules of the joint venture, to complete assignments and to enhance their efficiency ("The Regulations on Labor Management" 1980).

Joint ventures are also expected to cover medical costs for the period of an employee's contract. One can either contract with key hospitals or take care of employees through an employer's clinic. Medical insurance is the least favored system, with its high premiums and relatively low percentage of coverage. Joint ventures are also required by law to contribute 15 to 20 per cent of total wages to a housing fund for Chinese employees or to offer rental subsidies. Joint ventures should provide a retirement allowance, the rate of which ranges from about 20 to 30 per cent depending on the area. Joint ventures are required to make a payment of 1 per cent of the average salary of the area each month to the local labor institutions (Casati 1991: 16–22).

It is important to create a corporate culture that contains some Western management and is, at the same time, acceptable to the Chinese staff. Expatriate staff should try to serve as good role models. To help reduce the influence of *guanxi* (connection) and the Chinese reluctance to shoulder responsibility, several rules should be practiced. For example, detailed job descriptions should be provided, though it is impossible to cover every aspect of employees' responsibilities in a job description. Also, training should be integrated with daily operations. A corporate culture can be best established through close contact between expatriates and Chinese teams from the very beginning. A good nurturing environment for risk taking cannot be established without a safety mechanism of checks and balances for major decisions. People should also be allowed to make mistakes, as long as these do not threaten the normal operations and the safety of the staff.

SUMMARY AND CONCLUSION

This chapter has reviewed the major steps toward establishing a successful joint venture in China, which include finding good partners and locations; getting well prepared for China's legal framework for joint ventures; covering key issues in negotiations; taking effective measures to ensure a good foreign exchange balance and effective localization; and overcoming the difficulties in managing Chinese employees. As technology transfer is an integral part of many manufacturing joint ventures in the country, specific challenges pertaining to technology transfer are discussed. For many foreign investors, joint venture is probably the best if not the only option for them to enter the Chinese market. For others, a well-managed joint venture may significantly enhance their competitive advantages in China. Nonetheless, joint venturing in China is not an easy process.

In spite of all the well-known stories on the hardships of joint venturing in China, a recent survey shows that most joint ventures are making money, with 60 per cent reporting a return on investment (ROI) of 10 per cent or higher. Nearly a third reported an ROI equal to, or exceeding, 18 per cent. Although such returns would be considered minimal by venture capitalists in the United States, the average ROI of 11.6 per cent is respectable for China-based international businesses; only 12 per cent of the sampled firms appear to be losing money (Stelzer *et al.* 1991: 54). When more US companies do their homework and are well prepared for ventures in China, a higher percentage of success can be expected. Since 1990, the Chinese government has considerably enhanced transparency in its investment policies and promulgated a series of laws and regulations. Attracting foreign investment has become part and parcel of its long-term development strategy. Ventures involving the transfer of technologies, especially high technologies and those prioritized in the development program, will be particularly welcomed in the years to come.

16 Issues in dispute resolution

INTRODUCTION

Transactions of international technology transfer often involve long distances and people of diverse cultural and social backgrounds who may follow significantly different ways of managing technology transfer and its related businesses. This intricacy is further complicated by the fact that most technology transfer transactions require long-term relationships between the supplier and the recipient of technologies and involve complicated processes. These complexities considerably increase the chance that the parties to a transaction will be entangled in a contractual dispute at a certain point in their relationship. Even transactions negotiated with good intentions between familiar and friendly partners cannot exclude the possibility of serious conflicts. Thus, international managers should be fully prepared to handle such conflicts.

There are a number of ways to resolve business disputes, chiefly, litigation, mediation and arbitration, with each having some advantages and disadvantages. The selection of a proper way to deal with a specific business dispute is crucially important for the overall resolution. This is particularly so in transactions of international technology transfer, as both parties have to think of the undesirable consequence of losing economic rents if their cooperative relationship is abruptly cut short. This chapter introduces some of the most fundamental issues in dispute resolution, ranging from the legal environment of international litigation and key procedural issues in litigation (jurisdiction, venue and enforcement of international litigation) to alternative dispute resolution (ADR).

INTERNATIONAL LITIGATION

Disputes are not uncommon throughout the life of an international contractual relationship. Under most circumstances, however, the parties can privately achieve a mutually acceptable settlement. In those cases where a peaceful solution cannot be achieved, litigation may have to be considered. Usually, litigation is not among the most favored dispute resolution options for several reasons. First of all, litigation costs, both in terms of time and money, tend to be very high. Moreover, litigating international contractual disputes is very complicated and frequently ineffective due to the fact that no international court system has so far been established to resolve commercial disputes between private parties. Furthermore, lawsuits often wreak irreparable damage on business relationships, annihilating the possibility of future cooperation. Finally, litigation can lead to unwanted publicity. Having considered these perils of litigation, a company must carefully assess the options available for dispute resolution. Many firms choose alternative dispute resolution (ADR) as an option or in conjunction with litigation (Ratliff 1994: 9–10).

Legal environment of international litigation

In many cultures, litigation as a way to resolve business disputes has traditionally been discouraged. In traditional Chinese culture, for example, if parties are forced to settle a dispute either in the courts or through arbitration, they may feel that they almost openly admit they are not mature and sophisticated enough to manage their relationship. Consequently, if a conflict becomes publicly known, both parties will share the blame in the eyes of the disinterested Chinese, irrespective of the merit of any one party's position. This aversion to litigation is based on the Confucian priority of maintaining social harmony. In a similar vein, the Japanese have also encouraged alternative dispute resolution as a better way to resolve controversies than litigation (Eisenstadt and Ben-Ari 1990).

There is not yet a uniform set of rules or procedures governing contractual relationships between private parties. The body of international rules and procedures that is called international law (also referred to as public international law) is created to regulate relations between sovereign nations (Von Glahn 1992). The World Court (International Court of Justice), located in The Hague, only processes suits brought by sovereign nations, even though it handles commercial issues. When commercial disputes occur between sovereign nations, they may petition the World Court. Occasionally, a private party (company or individual) may request its home government to lodge its private claim against the host nation before the World Court (Fitzmaurice 1986).

Private parties have to bring their lawsuits before the domestic court systems. Their litigation belongs to private international law, which consists of the rules and procedures of national legal systems for resolving disputes involving litigants from different countries. These rules and procedures determine which nation's laws will govern the dispute, in which country or countries a lawsuit may be brought, how evidence may be gathered, and under what circumstances the courts of one nation will recognize and enforce a ruling made by the courts of another nation. When the disputed contract does not specify which body of law should govern the dispute, it may be determined by the rules of the forum where the lawsuit occurs. The process by which a court selects the appropriate law is known as conflict of laws (Collier 1987: 12–62). Most nations have established conflict of laws mechanisms.

Conflict of laws determinations in civil law systems normally can be achieved through the application of each nation's codified procedures for choosing the governing law. Common law systems basically follow one of two main approaches. According to the first and simpler approach, the governing law should be that of the place where the contract is signed, if the dispute concerns the formation of the agreement. But if the conflict focuses on the performance of the contract, the applicable law is usually that of the place where performance occurs. The second and more popular approach selects the governing law based on the most important contacts. The examination covers: the place of negotiation and contracting; the place of performance; the place of the contractual subject matter; and the place of incorporation, place of business and the domicile of the parties (Richards 1994: 203–204).

Jurisdiction

The most significant limit in international law on the right of nations to control business activities is probably the concept of jurisdiction (Neale 1988). Jurisdiction is basically embodied in two forms: power over the subject matter of a dispute and power over the individuals involved in the dispute. Most international business disputes are focused on the issue of personal jurisdiction. Therefore, a manager should be constantly vigilant as to when he/she might be dragged into court in a foreign country. Most nations maintain some form of "long-arm jurisdiction," allowing their courts to hear cases arising out of transactions occurring in the country involving a defendant outside the country.

Many countries have established specialized courts that have exclusive responsibility for certain commercial disputes. In the federal district courts of the United States, for example, a dispute between a US citizen and a foreign citizen must involve a claim of over $50,000 (Richards 1994: 198). Under the laws of some countries, failure to file a lawsuit with the court that commands proper subject matter jurisdiction may result in the plaintiff losing his/her cause of action. Even if the plaintiff is allowed to refile his/her lawsuit in the appropriate court, a significant loss of time and money has usually been incurred.

A court cannot handle a case unless it also has personal jurisdiction over the defendant. When the plaintiff files a lawsuit against a non-resident of the country, the court does not have personal jurisdiction unless the non-resident defendant has been proven to have certain minimum contacts with the country where the suit is brought. According to the rules of many countries, the minimum contacts requirement is met if the defendant intentionally does business there. The French Civil Code advocates a more controversial notion of personal jurisdiction for contractual disputes between French citizens and foreign non-residents. French courts will hear lawsuits brought by French citizens against foreign defendants, even though the transaction is made outside France by a French citizen who does not reside in France (Hotchkiss 1994: 34).

Many multinational corporations, concerned about jurisdiction over corporate parents, try to protect themselves against the personal jurisdiction of the courts in the foreign countries where they have business operations. The focus is laid on carefully defining relationships with a wholly owned subsidiary. Nevertheless, the foreign court may have personal jurisdiction over the parent if the corporate parent becomes too closely involved with the ultimate clients or if the subsidiary is undercapitalized, with inadequate assets to pay expected claims against it, or if the subsidiary is the *alter ego* of its parent, sharing the same officers and directors (Richards 1994: 199).

A defendant normally cannot be subject to the personal jurisdiction of a court if he or she has not received service of process. The issue of jurisdiction can become much more complicated when service of process is to be provided across national borders. When the defendant is physically situated in a foreign territory, some minimum contacts with the country of personal jurisdiction must be established. Moreover, international service of process can be both costly and difficult. Many courts require plaintiffs to comply with the cumbersome requirements of the 1965 Hague Service Convention and prohibit service of process on a foreign person by registered mail (Handbook 1983).

The Hague Service Convention is an international treaty created to facilitate the service of process across national borders and to provide proof that timely notice has been delivered. The United States and over thirty other nations have ratified the treaty (Handbook 1983). To serve process on a defendant in a member nation, a plaintiff may:

1 Follow any procedures established in the target nation to petition the institution concerned or the judicial personnel of the target nation to make the notification on the defendant.
2 Ask the home country's consular officers stationed in the target country to serve the process.
3 Comply with any procedures established by a treaty between the foreign nation and the nation where the court is located.

Collecting the necessary evidence in a foreign country is also a very costly and difficult part of international litigation. The court systems of some countries, such as that of the United States, attach great importance to "discovery" or "the pre-trial gathering of evidence" to eschew unfair surprises. To expedite the discovery process involving foreign parties, the United States and many Western European countries have ratified the Hague Evidence Convention. The foreign courts will perform the discovery process, if the request does not violate their national laws. But many other countries have not ratified this Convention nor do they have such broad discovery rules as the United States. Their courts disfavor the United States' exhaustive and burdensome discovery procedures. A US court's attempt to pursue its discovery process in Japan, for example, may not receive the cooperation of the Japanese system (Schaffer *et al.* 1993: 221–222).

Venue

The term "venue," sometimes confused with jurisdiction, refers to the appropriate geographical location of the court with jurisdiction. For instance, several federal district courts in the United States may have jurisdiction over a lawsuit, but only one of them is suitable as the proper venue (Schaffer *et al.* 1993: 211). Therefore, it is very important to determine which court is the proper place for the lawsuit to be brought. In a contract involving companies from two nations, the venue is often proper in both the supplier's and the recipient's countries. If the contract is formed in a third country and implemented in a fourth, courts in at least four nations could have proper venue.

Because the laws of the forum determine which body of substantive rules governs the dispute, the location where the lawsuit is brought can be crucial to the outcome. When a contractual dispute takes place, each of the parties opts for the court most advantageous to it. Therefore, parties to the dispute often question the suitability of a particular forum. Sometimes, two separate courts might be petitioned to simultaneously handle the same dispute, because one party to the dispute may dislike the current forum and file its own suit in a forum more favorable to itself. When such parallel proceedings in different jurisdictions take place, the first court judgment generally has a binding effect on the other court.

Sometimes, when a court meets both jurisdictional and venue requirements, a judge may dismiss the case on the basis of *forum non conveniens*. This practice is not usual outside the common law systems. In the United States, *forum non conveniens* has been increasingly utilized by MNCs to block foreign victims' attempts to take advantage of the liberal tort and product liability laws in the country. In determining the applicability of *forum non conveniens*, US courts usually adopt a three-step process. First, the court examines the availability of any "alternative forums" with subject matter jurisdiction over the dispute and personal jurisdiction over the defendant. Second, the court assesses the impact of various "private factors" on a fair trial, which comprise the costs of summoning witnesses, the ease of access to evidence and the enforceability of the judgment. Third, the court weighs certain "public factors," including the amount of court congestion, the intention to avoid burdening jurors with irrelevant issues or using a forum where the local community has a strong interest in the case, and the desire to have the case tried in a forum familiar with the governing law (Richards 1994: 202).

The Bhopal tragedy of India illustrates the usefulness of *forum non conveniens* for the MNCs (Schaffer *et al.* 1993: 211–212). The Union Carbide subsidiary was owned 50.9 per cent by the parent company, which had its headquarters in Danbury, Connecticut, while the Indians who controlled the plant owned 49.1 per cent. Lawyers representing Indian victims filed suit in the Federal District Court for the Southern District of New York, the district covering Manhattan. The judge dismissed the case on the doctrine of *forum non conveniens* and ruled that the case should be tried in India. The judgment was based on factors such as the cost of transporting witnesses and translation difficulties. Moreover, the judge stated that "to retain litigation in this forum (the United States) would be another example of imperialism, another situation in which an established sovereign inflicted its rules, its standards and values on a developing nation." The dispute was eventually settled in India in 1989 prior to trial, with the Union Carbide parent undertaking to provide compensation of $470 million.

Enforcement

When a lawsuit is concluded, the losing party is required to comply with the court decision. Although many parties voluntarily abide by the court decision, some may not obey the court. Consequently, the winning party may have to request a judicial order of enforcement. If the losing party physically resides or owns properties in the territory of the court's jurisdiction, it may not be so difficult to compel compliance. Nevertheless, if the losing party and its assets are located outside that jurisdiction, the winning party will have to enlist help from a foreign court to enforce the judgment. Not uncommonly, the foreign court neither enforces nor recognizes the judgment, leaving the plaintiff with no options other than more litigation.

There is no international institution to enforce judgments of the courts of sovereign nations. At the regional level, EC-EFTA countries took the lead in foreign enforcement by signing a Convention on Jurisdiction and Enforcement of Judgments on Civil and Commercial Matters in 1988. Some developing countries, such as Indonesia, do not enforce most foreign judgments. Other countries, like Norway and Saudi Arabia, only enforce foreign judgments from a country with which they have an enforcement treaty. Still others, such as South Korea and Austria, require reciprocity. Some Islamic countries do not enforce the interest portion of a foreign money judgment. Japan's courts only enforce civil judgments, not criminal ones. In the United States, courts enforce a foreign judgment when that judgment does not violate any strong public policy. Most American courts base this practice on the "doctrine of comity," which means the voluntary deference the courts of one nation pay to those of another (Richards 1994: 205–206; Schaffer *et al.* 1993: 228–229).

The most commonly used excuse for refusing to enforce a foreign judgment is that the original court did not have subject matter or personal jurisdiction. Some countries, like France and Brazil, vehemently uphold the principle of exclusive jurisdiction over their own nationals unless they themselves willingly accepted the jurisdiction of the foreign court. Other countries, such as India and the Philippines, refuse to enforce a judgment if their courts conclude that the foreign courts applied the inappropriate governing law. Many countries base their refusals to enforce foreign judgments on a wide range of public policy concerns. Greece and Thailand, for example, reject any judgments from a foreign court unless the plaintiff had a claim that would have been actionable in their country (Richards 1994: 205).

ALTERNATIVE DISPUTE RESOLUTIONS (ADR): MEDIATION AND ARBITRATION

ADR is often referred to as a good compromise since it gives disputing parties the opportunity to tell their stories to a neutral party and offers a way to resolve disputes promptly, confidentially and cost effectively. The two most commonly used methods for ADR are mediation and arbitration. Given the fact that the chances of resolving a dispute quickly and inexpensively are greatest with mediation, it should have been applied much more frequently than it has been up to now. Mediation also provides the parties with the opportunity to privately discuss settlement in a structured environment. Unlike mediation, arbitration has a binding effect upon the parties. The parties present their evidence to an arbitrator or panel of arbitrators who will judge the case at a hearing which is structured like a court trial. Mixed forms of ADR comprise mediation/arbitration, "baseball arbitration" (although this has been used primarily to resolve salary disputes in the world of professional sports, it has also been adopted in other areas), agreements concerning the amount of damages (e.g. a claimant may agree to a damage ceiling in exchange for a minimum award) and minitrials (Berman 1994: 76–77). In view of the necessity of developing long-term cooperative relationships in technology transfer, ADR is particularly attractive to the supplier and recipient of technologies.

Mediation

As an increasing number of courts require the implementation of ADR prior to the pursuit of litigation, mediation is becoming more popular (Nolan-Haley 1992). The Commercial Court in London, for example, has directed the Official Referees to suspend court proceedings until after parties try to settle their disputes by alternative means. The number of private mediation firms has also grown. The range of disputes they manage includes such issues as general commercial contract, partnership, investment, construction, professional indemnity, insurance matters, charity and trade. As a result of a growing number of transnational joint ventures and strategic alliances, mediation is becoming very valuable in multiparty disputes.

Major mediation techniques in commercial disputes can be briefly summarized in four stages (Nolan-Haley 1992: 60–77):

1. Preparation of a short written summary of the case for the mediator.
2. An initial joint meeting for oral presentations between the parties.
3. Private meetings that the mediator then has with the parties to clarify issues and search for settlement possibilities.
4. A series of joint meetings which are called to continue the search for solutions, sign the agreement or conclude the mediation.

However, there still exists considerable apprehension about the use of mediation. Financial institutions, for example, have not traditionally pursued mediation to any significant extent. Reasons for the apprehension are based on the fear that the mediator would not be much better qualified in the complexities of a financial dispute than a judicial authority. There is also a lack of trust in the method of this conflict management approach or with the professionals involved in it. Moreover, many believe that the mechanics of the informal process would not be adequate for the "extensive factual, technical and financial material" associated with banking matters (Barnett 1994).

Arbitration

Many countries have bilateral treaties that provide for arbitration of disputes involving at least one sovereign nation. The International Center for the Settlement of Investment Disputes (ICSID) was created by the World Bank to promote mutual trust between private foreign investors and the host governments. So far, more than ninety nations are party to the ICSID Convention. As part of the World Bank, ICSID provides a forum and administration for the reconciliation and arbitration of investment disputes (Fox 1992: 296–299). Arbitration has many advantages: it is normally faster than litigation, is conducted privately, can involve non-lawyers as arbitrators and has a better chance of being enforced than a foreign court's judgment. Consequently, international contracts often designate arbitration as the exclusive form of dispute resolution (Redfern and Hunter 1986).

A recent example of major MNCs choosing arbitration involved

IBM and Fujitsu, which selected arbitration to settle their dispute even after litigation had already begun (Schaffer *et al.* 1993: 231). Each party selected an arbitrator, one of them being a computer specialist and the other a law professor from Stanford University. The focus of the dispute was the operating system software for IBM computers, and the alleged Fujitsu infringement on IBM's rights. Fujitsu contended that IBM was engaging in legal maneuvers to frustrate Fujitsu's competition. Because the law concerned was not clear, both parties agreed to arbitration. On September 15, 1987, the two companies settled a four-year legal battle by the decision of two arbitrators from the American Arbitration Association, who awarded IBM $833.2 million and gave Fujitsu access to the operating system without fear of future suits. The IBM–Fujitsu case presented a unique solution to a highly intricate dispute.

However, businesses cannot be forced to submit their contractual disputes to arbitration. The disputing parties must voluntarily agree to the use of an arbitral forum. In order for arbitration to be effective, the disputants should explicitly spell out in their business contract the conditions under which arbitration will take place. The arbitration clause should also clearly refer to the procedural rules of the chosen sponsoring institution. One major challenge is to include an arbitration clause that is complete enough to cover a variety of circumstances and factors, yet simple enough that the other party will accept it quickly without excessive negotiation (Mason 1994: 22–26).

As in litigation, arbitration is subject to cultural influences. For instance, the Japan Commercial Arbitration Association (JCAA), which was established to handle disputes in international business, based its rules on the model of the American Arbitration Association (AAA). Despite similar arbitral rules, the United States and Japan are reluctant to utilize each other's commercial arbitration association procedures, because their approaches toward dispute resolution are different. While the basic procedural rules of the JCAA and AAA share common denominators, the distinctions arising from subtle variations and incongruous interpretations are not insignificant. Under Japanese judicial guidance, the JCAA has established a non-adversarial arbitration process. As a result, the Japanese arbitration mechanism represents a blend of conciliation and arbitration, distancing itself from litigation as much as possible (Hanlon 1991: 603–626).

Well-known institutions for international arbitration

Most countries in the world have institutions responsible for commercial arbitration. For cases involving international arbitration, a few institutions are most frequently used and their rulings are quite influential. The following is a brief introduction to these institutions.

International Chamber of Commerce (ICC)

Headquartered in Paris, the ICC has regional offices all over the world. The ICC hears disputes in any of these locations under whatever law the parties concerned choose (Craig *et al.* 1985). Currently, the ICC is the best known and probably the most frequently used of the arbitral institutions in the world. In performing arbitration, the ICC follows several fundamental principles. These include rules that: a decision must be made within six months after the hearing closes unless an extension is granted; the right to present evidence and call on witnesses lies with the arbitrators, though the parties may require that decisions be exclusively based on documentary evidence; and arbitrators may consider custom and trade usage in deciding the case.

The World Intellectual Property Organization (WIPO) Arbitration Center (WIPOAC)

Established on July 1, 1994, WIPOAC is believed to be uniquely placed to play a positive role in the resolution of international intellectual property disputes. The Center's mission in the management of proceedings is to provide maximum administrative and organizational support for arbitrators to efficiently and effectively discharge their duties. Facilitating the resolution of such disputes is also expected to improve WIPO's efforts to promote the protection of intellectual property. The Center constitutes an administrative unit of the International Bureau of WIPO. Its activities are supervised by a Supervisory Board on Arbitration, which comprises representatives drawn from both the public and private sectors (Gurry 1994: 4–6).

American Arbitration Association (AAA)

With its headquarters in New York City, the AAA handles both domestic and international disputes. The AAA's rules are relatively more detailed than those of the ICC and closer in concept and scope to the American litigation system (Fox 1992: 256–258). Apart from providing administrative support for the proceedings, the AAA can also designate the actual arbitrators upon the request of the disputants. A few fundamental principles provide guidance to the AAA's arbitration, including rules that: a decision should be made within thirty days after the hearing closes; the parties may present evidence and call on witnesses, but they may also agree that the arbitrators' decisions will be based on documents only; and arbitrators may consider custom and trade usage.

London Court of International Arbitration (LCIA)

One outstanding feature of the procedural rules of the LCIA is that they generally provide more details than other major international arbitral institutions. Another unique feature of the LCIA is the traditional support of English courts, which readily enforce an arbitrator's discovery orders and remove an arbitrator who refuses to proceed in a timely manner. According to the rules of the LCIA, the arbitrators must make a decision as soon as practicable. The disputants may present evidence and require that the final decision be only based on documents, but the arbitrators have the right to prevent the cross-examination of witnesses. There is no specific provision for custom and trade usage, and the arbitrators might not consider them in reaching the final decision (Richards 1994: 213).

Other institutions and ad hoc *arbitration*

Other major arbitrary institutions include the Zurich Chamber of Commerce, the Japan Commercial Arbitration Association and the Stockholm Chamber of Commerce. In recent years, other centers of commerce, such as Brussels, Hong Kong, Cairo and Kuala Lumpur, have begun to emerge as sites for international commercial arbitration. When seeking sites for arbitration, people tend to consider such factors as good telecommunications, ease of access, availability of experienced arbitrators and a non-hostile judiciary.

In addition, the parties may proceed completely on their own under a form of dispute resolution known as *ad hoc* arbitration. Under this approach, the parties themselves work out the process and administration of the arbitration. They provide for selection of the arbitrators, the site of the arbitration, the language of the arbitration, the rules applicable to the proceedings and applicable law. The administrative expenses and arbitrators' fees and costs are resolved among the disputants themselves. Major advantages of this approach include lower costs and fees, complete control by the parties over the whole process of the arbitration and a generally faster process (Fox 1992: 249–252).

UNCITRAL arbitration procedures and the New York Convention of Enforcement

Comprehensive, generally applicable and widely accepted, UNCITRAL procedures for international arbitration have been promulgated by the United Nations Commission on International Trade Law for the purpose of accommodating the differences and similarities among the world's major legal systems (Dore 1986). The UNCITRAL procedures are designed to work among businesses from all over the world and to help reduce the differences of culture, language and legal systems that frequently obstruct dispute resolution. Although the UNCITRAL arbitration procedures were initially created for *ad hoc* arbitration, most formal arbitral institutions will comply with them if the parties require those procedures in their agreements. The UNCITRAL arbitration procedures include the following rules: no time limit is set for issuing a decision; the arbitrators must consider custom and trade usage; the arbitrators have considerable discretion in deciding the admissibility of evidence and of witnesses; and the losing party should bear all the relevant costs of the proceedings.

In 1958, the United Nations drafted the Convention on the Recognition and Enforcement of Foreign Arbitral Awards. This treaty, also known as the New York Convention, provides for the recognition of arbitration agreements and for the enforcement of foreign arbitral awards (Van den Berg 1981). As of January 1, 1991, less than eighty nations have ratified the New York Convention. The Convention requires each country to recognize and enforce arbitral awards based on agreements "in writing" which "include an arbitral clause in a contract or arbitration agreement, signed by the parties or contained in an exchange of letters or telegrams." There are very few cases in which a contracting country may refuse recognition and enforcement. Refusal situations occur when:

1 The recognition or enforcement of the award would violate the public policy of that country.
2 The subject matter of the award is incapable of settlement by arbitration under the law of that country.

3 The award covers a difference not contemplated by or not falling within the terms of the submission to arbitration.
4 The party against whom the award is made was not granted appropriate notice of the appointment of the arbitrator or of the arbitration proceedings.
5 The composition of the arbitral authority or arbitral procedure was not in conformity with the agreement of the parties.
6 The agreement is not valid under the law to which the parties have subjected it or under the law of the country in which the award was made.

Although almost half of the UN member nations have not yet accepted the Convention, arbitration can still be enforced through various kinds of bilateral agreements of the countries in question or their internal arbitration systems. With the rapid expansion of international trade and investment, arbitration is becoming an increasingly accepted way to resolve disputes all over the world.

SUMMARY AND CONCLUSION

This chapter has discussed some fundamental issues of international litigation as well as alternative dispute resolution. Because international technology transfer transactions often involve complicated processes and have an extended duration, the chance that the parties to a transaction may become involved in a dispute is fairly high. As the domestic court systems of the countries in the world vary considerably, litigation of private international disputes can be a confusing, costly and frustrating experience. Moreover, by resorting to judicial remedies, the parties risk severely damaging or completely destroying what could still be a profitable relationship. Frequently, a domestic court is reluctant to enforce the judgments of a foreign court, thus forcing the litigant either to go through the expensive process of further litigation or to totally abandon the case. Thus, many international managers have tried to settle their disputes via mediation and arbitration.

Major steps can also be taken to reduce the unnecessary complexities of litigation at the stage of contract formation. To forestall confusion in a potential litigation, parties can insert a forum selection clause in their contract, which specifies where litigation is to be located. Contracting parties can also significantly reduce uncertainty by placing a choice of law clause in their agreement. As a common practice, the courts in most of the world's major trading countries enforce such a provision as long as some sort of connection is established between the chosen legal system and the contract. International managers should also pay attention to the potential impact of cultural differences upon litigation. As mentioned earlier, many cultures do not encourage litigation nor are they committed to its rigid enforcement. To deal with such a complication, managers may find it helpful to formulate a flexible agreement that anticipates future conflicts and provides a feasible mechanism for renegotiation.

Concluding remarks

INTRODUCTION

In this book, international technology transfer has been examined by utilizing conceptual ideas from A. C. Samli's general model, which comprises five key components: the sender, the technology, the receiver, the aftermath and the assessment (Samli 1985: 3–15). The main focuses of the book have been on four major dimensions: international technology transfer and comparative governmental policies; international technology transfer and international protection of intellectual property; international technology transfer in the form of licensing; and international technology transfer through a few major commercial channels. The previous sixteen chapters have provided abundant evidence that international technology transfer is an interdisciplinary subject, deserving a comprehensive approach on research.

Technology transfer, already difficult in a purely domestic environment, is much more so in the international context. Differences in political and economic systems, development strategies, national cultures and legal structures create hindrances to transnational technology transfer that are not found in domestic cases. As of now, international technology transfer is still a relatively young discipline, with many problems to be solved and many questions to be answered. However, international technology transfer is also a discipline with a prospect for dynamic growth in the years to come. This concluding chapter discusses the dynamic relationships between international technology transfer and international business, assesses the overall costs and benefits of international technology transfer, analyzes major existing problems in this field of study and finally makes a few suggestions for future improvement.

INTERNATIONAL TECHNOLOGY TRANSFER AND INTERNATIONAL BUSINESS

There is no process of international business without technology transfer, particularly the process-embodied transfer. With the increasing globalization of international business, technology-intensive industrial sectors, such as commercial aircraft production or semiconductor computer chip manufacturing, have become a major part of international business. American managers are currently learning what many of their European and Asian counterparts have known for some time: "To survive in the future a firm, especially a multinational corporation (MNC), must be able to transfer technology to other countries better than the competition" (Keller and Chinta 1990: 33).

Because the main focus in the study and the practice of international technology transfer is on the transfer that occurs across national borders, understanding the fundamental operations and environment of international business is indispensable for those who are engaged in the study and the practice of technology transfer. On the other hand, the study of international business is not complete if international technology transfer is not included. In fact, technology is the most significant competitive advantage which MNCs have to develop and maintain in order to increase their market share in the world. Most current international business textbooks have devoted a section to the discussion of international technology transfer (Asheghian and Ebrahimi 1990: 289–314; Grosse and Kujawa 1995: 500–520, etc.).

Unlike many subjects of study, international business has so far not enjoyed a distinct definition. The traditional approach to international business has evolved from an extension of the functional fields of management. For this reason, some writers (Robock and Simmonds 1989: 3) have defined it as "a field of management training that deals with the special features of business activities that cross national boundaries." A more comprehensive understanding of international business has been gradually developing from the fields of international marketing, finance, business operations and human resources management. However, the functional areas of management studies seldom overlap in business schools, and the interdisciplinary nature of international business contributes to the lack of consensus concerning the domain of international business.

To break away from this dilemma, Toyne (1989: 1) argues that the confusion is mainly caused by a misplaced emphasis on the firm-as-a-unit analysis. Therefore, if exchange were made the unit of analysis, it would be possible to define the field on a more feasible basis. For Toyne, exchange constitutes the core of international business. Along this line, Vaghefi, Paulson and Tomlinson (1991: 4) define international business as "a dynamic field of activity involving an extensive exchange process that encompasses not only goods but also social, cultural, and human interactions across national boundaries." The international technology transfer that has been discussed thus far in the book is obviously a component part of this exchange process.

There should be no surprise that international technology transfer has been studied from different angles by different groups of scholars. First, it has been widely researched by international study specialists, with a focus on the role of governments and their policies and the interrelationship between international technology transfer and development issues. However, there has been little attempt by these scholars to establish a linkage between management and international technology transfer. Second, many management scholars have done research through case studies that are not devoted to a discipline or a functional field. They often have to utilize more than one functional field of management, but there has been no attempt to bridge the gap between international studies and management. Third, many publications in the field have been written by law professors and lawyers. In general, they are more concerned with the legal implications of international technology transfer transactions, such as contract formation, intellectual property protection and dispute resolution.

In this book, the three fields of international studies, international management and international law have been organically integrated. This has been made possible by a multidimensional approach and a complex structure of the four parts of very different dimensions. While each part has its own focus, they are designed to be mutually complementary and organically linked to each other. There is a cohesive chain of topics throughout the text. The chain begins with governmental policies on international technology transfer and investments (the fundamentals of the operational environment of technology transfer), proceeds to international protection of intellectual property (the basis of technology transfer), then discusses various aspects of international licensing (the traditional core of technology transfer) and ends with a general introduction to other commercial channels of international technology transfer (the multiplying areas of technology transfer), with a focus on two of the most common channels, international subcontracting and joint ventures.

The close integration of international management, international studies and international law under international technology transfer is very important to the development of the discipline of international technology transfer. As Agmon and Von Glinow have succinctly summarized:

> A better understanding of the nature of the process of international technology transfer will make those researchers and practitioners who are dealing with issues of international business better equipped to carry out their mission. Understanding the complex environment of international business, and the realization that technology transfer is a necessary, and integral part of international business may bring technology transfers to the center stage of international management where it belongs.
>
> (Agmon and Von Glinow 1991: 273)

COOPERATION AND CONFLICTS AMONG THE MAJOR PLAYERS OF INTERNATIONAL TECHNOLOGY TRANSFER

Technology transfer has dramatically revolutionalized international trade and greatly increased the involvement of different countries in the flow of goods and services across national boundaries. In addition to general economic benefits to their respective home societies in terms of export promotion, increased job opportunities and technology advancement, technology transfer transactions also generate economic rents that both the transferor and transferee may share. However, like other forms of international business, technology transfer not only brings specific benefits but also some costs to the participants. As has been discussed throughout the book and by many other scholars (e.g. Aggarwal 1991: 68–70), those benefits and costs vary a great deal for transferors and transferees, depending upon the channels selected for the transfer.

As is stressed repeatedly throughout this book and in the publications of some other scholars, technology transfer, by its nature, should not be a "win–lose" situation or what is known as "zero sum" game. Rather it should be conducted in accordance with the principle of a "win–win" or "cooperative" game. The incremental value created by a technology transfer transaction constitutes the basis of the economic rents that both the transferor and the transferee may share. If they fail to cooperate successfully, the economic rents they share will either be smaller than possible or disappear completely. It is not merely an issue of the transferor maximizing its share of earnings at the expense of the transferee. "More fundamentally, the issue is how the total economic value can be maximized in the first place in the licensee's (transferee's) territory, with the question of sharing the gains as a second-tier issue" (Contractor 1985: 199).

The successful conclusion of negotiations with the signing of a contract only marks the beginning of a continuing relationship between the transferor and transferee. As has been discussed earlier, transferring a technology from one company to another is rarely a one-time, single act, but rather an ongoing process. A technology transfer agreement should be a mutual commitment to work together for the benefit of both sides. Thus, a good technology transfer arrangement is a long-term, cooperative venture. It is not to the advantage of either the transferor or the transferee to strike a hard bargain that will cause bad feelings in the future (Root 1981: 128). In short, trust and a cooperative spirit are indispensable for an agreement to be viable in the long run.

However, technology transfer, by its nature, is a very complicated process, which may involve multiple players. Because technology normally does not have a clear-cut market value and the negotiation process is characterized by a bilateral monopoly, the bargaining process can be extremely intricate and difficult, with each participant trying very hard to increase its share of the economic rents. In order to establish a common ground for bargaining, the transferor and transferee have to close the gaps in their ceiling and floor price offers. This process is further complicated by some environment-specific factors, such as governmental regulations, political and business risks, levels of competition for the technologies and so on (Root and Contractor 1981: 23–32; Cho 1988: 70–79).

In addition to the complications of price, bargaining is another major uncertainty, i.e. the issue of appropriate technology which has always been controversial between the transferor and the transferee as well as their respective home countries. Technology is not universal. No single technology is appropriate for all purposes, and every technology is suitable for meeting some goals. Both the transferor and the transferee should pay attention to the appropriateness of the transfer (Heller 1985: 65–82). Technology appropriateness has both macro and micro dimensions. The macro dimensions comprise such issues as the impact on employment and shifts in the overall balance of power among the nations involved. The micro dimensions deal with the direct and indirect impacts upon the participants of technology transfer.

Technology developed in one society and transplanted to another without adaptation may be analogous to a living organ transplanted from one body to another without testing compatibility. If the transplanted organ is not compatible, it is highly likely to be rejected, thereby causing serious damage, even death, to its host (Robinson 1988: 214). For many in the transferee countries, the technology transferred by MNCs is often "inappropriate," because it does not utilize host country factors of production to the best advantage. There are many examples of inappropriate technology transfer, such as the transfer of labor-saving technology to a society suffering from under or unemployment, or the establishment of a manufacturing facility using potentially dangerous processes or products in a country lacking adequate inspection and enforcement procedures, or the transfer of technology that may cause severe environmental degradation.

From the perspective of the transferor countries, inappropriate technology transfer may pose a serious threat to their national security and severe challenges to the world competitiveness of their national economies. Out of concern for national security, Western countries imposed strict export controls against the former Eastern bloc countries via the Coordinating Committee for Multilateral Export Control (COCOM) during most of the postwar period. To preserve the competitiveness of national economies, many countries have created various policies and regulations aimed at keeping and developing high-tech industries in their home territories. Many in the transferor countries are vocal in their opposition to technology transfer because such transfer may result in the loss of the industrial base and jobs.

At the micro level, the transfer of inappropriate technology is a very subtle issue of balance for the transferee. State-of-the-art technology may prove both too costly and too difficult to use for some transferees and the transfer of out-dated technology or rapidly outdating technology represents a waste for most. For the transferor, the transfer of an inappropriate technology can prove to be a very costly and ineffective process. Another major concern is the possible creation of a competitor. The nature of bilateral monopoly in the transfer process has made this issue more complicated, as communication on technology is subject to various restrictions.

The third uncertainty lies in the very process of international technology transfer. As has been explained, technology can be transferred through various channels, with each having its own unique process. It is impossible to generalize a standard process of international technology transfer and different approaches on restrictive

Table C.1 Major actors and issues in the international process of technology transfer and development

Issues/actors	*Supplier firms (transnational corporations)*	*Recipient governments (less industrialized countries)*	*Recipient firms (technology) (recipients in LICs)*	*Supplier governments (homes of TNCs)*
1 Contribution	We do the best we can under each circumstance but local governments and firms are also responsible for maximizing the contribution of imported technology.	Imported technology must be appropriate, it must contribute to our technological growth, and to our society. We want growth, not dependence. Most technologies are inappropriate.	Technology we receive must strengthen us and make us self-sufficient. Sometimes we do not receive the support we expect from suppliers.	Impact of imported technology depends on host government policy and available capacity as well as on TNCs. TNCs alone should not be blamed for impact.
2 Control of technology	We must control our technology and how it is utilized, so we can maintain our competitive position and improve upon our technology. It is our most important asset.	We have to have control over imported technology, to improve our science and technology position. Lack of local control hinders our policies.	We don't feel as though technology belongs to us, so we don't make every effort to improve it. If we had full control, we would use technology better and would improve it and ourselves.	You cannot expect a private firm which has spent millions of dollars and resources on developing a technology to forfeit its right to control completely. Each recipient is but one of many. Suppliers need to maintain control.

3 Restrictive transfer practices	If we require safeguards or certain quality standards or have other requirements, it is only because we care about our quality, reputation, and the preservation of our markets and competitive position. We abide by the laws.	Restrictive and abusive practices of TNCs have drained our treasures while keeping us dependent and techologically backward. All such restrictions must be eliminated.	Some restrictions make it costly or uneconomical for us to have contracts for acquisition of foreign technology; but we need them to grow, make new and better products, sell more. Restrictions inhibit us	Restrictive practices which impede the functioning of the market are prohibited by our laws, which are sufficient to prevent such practices. But most of what recipients call restrictive are normal outcomes of bargaining situations.
4 Price	We have spent much time and money, and sell our technology in the market like any other item. If we don't get the right price we won't transfer our technology.	Technology suppliers must charge us the marginal cost of technology, as our markets were not the prime motive for the initial development of these technologies.	The price of technology is often too high for us to afford, but we have to pay what they ask.	Price is always subject to the laws of supply and demand. Buyers and sellers are mature and can agree without government interference.
5 Government control (inc tax, patent, . . .)	Too much government control hampers our operations, reduces our effectiveness.	We must have control over our society and the use of our productive resources.	We need freedom from unnecessary government interference, but we also need government protection.	Unless matters of national interest are involved, firms must remain as free as possible.

Table C.1 cont

Issues/actors	*Supplier firms (transnational corporations)*	*Recipient governments (less industrialized countries)*	*Recipient firms (technology) (recipients in LICs)*	*Supplier governments (homes of TNCs)*
6 Protection of proprietary rights	We need confidentiality and protection for our most important asset. Without it, there would be no incentive to develop new technology and to transfer it.	We need the information for strengthening our national technological capacity. TNCs are too restrictive and secretive about their technology.	We respect the desire of suppliers for confidentiality and abide by our contracts. But if a needed technology is too costly or unobtainable from the source, we will copy it if we can.	One cannot expect owners of technology to give away major assets.
7 Assessment	We will provide as much data as we can, but much of data sought is not available even to ourselves. Technology package cannot be unbundled without increased cost or damage to quality. How can you assess fifty years of R & D?	We often have to buy unnecessary technology just because it comes in the package. Often we don't know why we are paying. No clear data or criteria exist.	Often we don't understand why we have to pay certain costs or receive certain technologies which we can't use or don't need. But we can't shop around, have no resources to search.	Recipients often do not realize that the package is needed, and cost of technology cannot be readily assessed. It has taken many years to develop.

8 Conflict resolution	We prefer our own courts, but will settle in international tribunals. We seek impartial and non-partisan courts.	We have laws and courts and will resolve any conflict according to these laws. We cannot accept foreign interference in our sovereignty.	We follow our government's courts and laws.	Many LIC courts are partisan and many laws are unfair toward foreigners. Also subject to political pressures. We prefer international bipartisan arbitration or settlement in our courts. Our laws are often adequate to handle international disputes.

Source: Perlmutter and Sagafi-nejad 1981: 26–27

business practices have further complicated the issue. On the whole, the transfer process is not only determined by the nature of the technology transfer, but influenced by the strategies of both transferor and transferees and their relative competitive positions.

To better understand the relationships among the major players in international technology transfers, H. V. Perlmutter and Tagi Sagafi-nejad (1981: 3–31) have provided a compendious description of various issues that are involved between an MNC transferor, host government, the transferee local firm and the home government of the MNC transferor. Conflicts in objectives, strategies and final result have exerted and will continue to exert significant influence on the four players in the process of technology transfer.

TENTATIVE SUGGESTIONS

International technology transfer can be improved at two levels. At governmental level, national governments should improve communications between each other and cooperate in working out a sustainable and widely acceptable international code of technology transfer. In this regard, the United Nations and its specialized agencies, such as UNCTAD, UNIDO, the UN Multinational Corporation Center and WIPO, have done very useful work by promulgating various model laws applicable to international technology transfer, though they have not so far been accepted by many countries. A widely endorsed general code of technology transfer can greatly facilitate the flow of technology across national borders. In this direction, the effort by UNCTAD to establish an international code of conduct on the transfer of technology has made significant progress, but the developed and developing countries have not yet been able to resolve differences over the content of the restrictive business practice provisions of the code, applicable law and dispute settlements (Blakeney 1989: 59–161).

A fair and sustainable system of international property protection is a key to the greater flow of technology across national borders. International intellectual property protection has become a major issue in international economic relations. Although an increasing number of countries have joined various conventions and agreements on intellectual property protection, enforcement in the developing countries has remained a major problem. While developing countries do have some valid concerns on the protection of intellectual property, on balance they benefit more from strengthening intellectual property protection in the long run. Inadequate protection of intellectual property will not only undermine their normal exports to developed countries, but also thwart their own native innovative activities.

There has also been a proposal on the establishment of large technology data banks, to store the world's technology. Such banks are to provide information with regard to who has the capability to do what and the conditions one has to meet to access relevant technology. Enhanced knowledge on the part of potential technology buyers would make the international market for technology considerably more competitive. The access of developing countries' governments and companies to such data should be allowed at nominal cost. Several international organizations, including the United Nations, the European Union, the Andean Common Market and many private companies, have already made progress in this direction. So far, the Japanese general trading companies have maintained probably the largest data banks, though access to these banks can be quite expensive (Robinson 1988: 216).

To facilitate technology transfer across national boundaries, governments of major technology exporting countries are urged to reduce or waive domestic taxes levied on royalties and the fees derived from technology transferred to developing countries and on the salaries of their citizens whose main activity is technology transfer to developing countries. Greater flows of students, scholars and technicians from developing countries to developed countries should be facilitated with more support from developed countries, though they should not be encouraged to stay on a permanent basis. On the other hand, developing countries are urged to improve their own environment for technology transfer. They should, for example, strengthen their enforcement of awards rendered by foreign arbitration tribunals (Robinson 1988: 217).

At the educational level, the traditional curriculum of business schools should be changed to accommodate the need of international technology transfer by establishing disciplines in technology transfer and foreign direct investment management. Technology transfer is not merely a part of technology management, but includes marketing, organizational, investment and financial issues and certainly has political and social dimensions. Courses in this discipline should include comparative national technology and investment policy studies, economic and social development studies, business ethics, intellectual property protection, legal aspects of technology transfer and the management of technology transfer through various channels. Such curricula can help future managers deal with the complexities and challenges of international technology transfer. In addition, coordinated research efforts in international technology transfer by scholars, specialists and practitioners from the above relevant fields may afford better understanding of various uncertainties in the subject.

Improvement can also be made at the corporate level, where technology transfer is becoming an increasingly important advantage in global competition. Companies can use several intrafirm and interfirm bonds to facilitate the transfer of technology across organizational and international boundaries (Keller and Chinta 1990: 37–40). Intrafirm bonds can be provided via growth strategies (such as the expansion of existing products and services into foreign countries) and redeployment strategies, which encompass parallel, delayed and sequential introductions of existing products or processes into another country. Interfirm bonds include a wide range of forms, such as joint ventures, wholly owned foreign subsidiaries and licensing. Technological compatibilities can help form a bond and cultural familiarities between firms from different countries that have similar values can seal the bond for technology transfer. Appropriate organizational designs and competitive strategies as well as human resource management and organizational control policies should be developed to create and strengthen bonds for international technology transfer.

SUMMARY AND CONCLUSION

In earlier sections, the relationship between international technology transfer and international business is discussed. Various issues concerning technology transfer and the dynamic relationships among major players involved in technology transfer are also considered. At the end of the chapter, a few tentative suggestions are made to improve global technology transfer. At present, many of the important factors in international technology transfer remain difficult to measure and a large number of questions are left unanswered. Managerial challenges are daunting, with many issues remaining uncertain. Environments in many countries are still inhospitable, exposing the companies engaged in the business to high risks and high barriers.

Meanwhile, the need for international technology transfer is continually growing at a fast pace. Well-managed technology transfer should bring benefits to both transferors and transferees and their respective home societies. As Samli comments:

> If all nations, through technology transfer, benefit from technological advances, first, the gap among the rich and the poor will

narrow. Second, all nations, rich and poor, will learn to be more efficient and therefore there will be less waste in the world's industrial endeavors. More knowledge, better utilization of resources, fast industrial progress, and therefore, elimination of economic underdevelopment are all feasible outcomes of successful technology transfer.

(Samli 1985: 3)

References

INTRODUCTION

Aggarwal, Raj (1991) "Technology Transfer and Economic Growth: A Historical Perspective on Current Development," in Tamir Agmon and Mary Ann Von Glinow (eds) *Technology Transfer in International Business*, New York: Oxford University Press, pp. 56–76.

Betz, Frederick (1987) *Managing Technology: Competing Through New Ventures, Innovation, and Corporate Research*, Englewood Cliffs, NJ: Prentice-Hall Inc.

Chudson, W. A. (1971) *The International Transfer of Commercial Technology to Developing Countries*, New York: UNITAR.

Contractor, F. J. and Sagafi-nejad, T. (1981) "International Technology Transfer: Major Issues and Policy Responses," *Journal of International Business Studies* (Fall): 113–135.

Erdilek, Asim and Rapoport, Alan (1985) "Conceptual and Measurement Problems in International Technology Transfer: A Critical Analysis," in A. C. Samli (ed.) *Technology Transfer: Geographic, Economic, Cultural, and Technical Dimensions*, Westport, Conn.: Quorum Books, pp. 249–261.

Frankel, Ernest G. (1990) *Management of Technological Change*, Dordrecht/Boston: Kluwer Academic Publishers.

Goulet, Denis (1978) "Dynamics of International Technology Flows," *Technology Review* (May): 32–39.

Jolly, J.A. and Creighton, J.W. (1975) *Technology Transfer and Utilization*, Monterey, Calif.: Naval Postgraduate School.

Li, Shao-qing (1993) *International Technology Transfer*, Guangzhou: Jinan University.

McIntyre, John R. (1986) "Introduction: Critical Perspectives on International Technology Transfer," in John R. McIntyre and Daniel Papp (eds): *The Political Economy of International Technology Transfer*, Westport, Conn.: Quorum Books, pp. 3–24.

Rugman, Alan M. (1983) *Multinationals and Technology Transfer*, New York: Praeger.

Samli, A. C. (1985) "Technology Transfer: The General Model," in A. C. Samli (ed.) *Technology Transfer: Geographic, Economic, Cultural, and Technical Dimensions*, Westport, Conn.: Quorum Books, pp. 3–15.

Schnepp, Otto, Von Glinow, Mary Ann and Bhambri, Arvind (1990) *United States–China Technology Transfer*, Englewood Cliffs, NJ: Prentice-Hall.
Stewart, Charles T. and Nihei, Yasumitsu (1987) *Technology Transfer and Human Factors*, Lexington, Mass.: Lexington Books.
Vernon, Raymond (1986) "The Curious Character of the International Technology Market: An Economic Perspective," in John R. McIntyre and Daniel Papp (eds) *The Political Economy of International Technology Transfer*, Westport, Conn.: Quorum Books, pp. 41–57.
World Intellectual Property Organizations (WIPO) (1977) *Licensing Guide for Developing Countries*, Geneva: WIPO.

1 COMPARATIVE NATIONAL MODELS OF TECHNOLOGY ACQUISITION

Conroy, R. (1986) "China's Technology Import Policy," *Australian Journal of Chinese Affairs* 15: 22–24.
Danilov, Dan P. (1989) *Immigrating to the USA*, Vancouver, BC: Self-Counsel Press.
Evans, Peter (1979) *Dependent Development: The Alliance of Multinational, State, and Local Capital in Brazil*, Princeton, NJ: Princeton University Press.
Gardner, Robert W. and Bouvier, Leon F. (1990) "The United States," in W. J. Serow, C. B. Nam, D. F. Sly and R. H. Weller (eds) *Handbook on International Migration*, Westport, Conn.: Greenwood Press, pp. 341–362.
Hirschmeier, Johannes and Yui, Tsunehiko (1981) *The Development of Japanese Business 1600–1980*, London: George Allen & Unwin.
Jasny, Naum (1961) *Soviet Industrialization, 1928–1952*, Chicago: University of Chicago Press.
Jones, Maldwyn A. (1992) *American Immigration*, Chicago: University of Chicago Press.
Li, Debin (1985) "Questions of Our Import of Technology and Equipment in the 1950s," *Beijing Daxue Xuebao* (Beijing University Journal) 4: 78–85.
Mandi, P. (1981) "The Brain Drain – A Sub-system of Center-Periphery Relationship," *Development and Peace* 2: 35–52.
Mirza, Hafiz (1986) *Multinationals and the Growth of the Singapore Economy*, New York: St Martin's Press.
Nove, Alec (1984) *An Economic History of the USSR*, New York: Penguin Books.
Nussbaum, Bruce (1983) *The World After Oil: The Shifting Axis of Power and Wealth*, New York: Simon & Schuster.
Ozawa, Terutomo (1985) "Macroeconomic Factors Affecting Japan's Technology Inflows and Outflows: The Postwar Experience," in Nathan Rosenberg and Claudio Frischtak (eds) *International Technology Transfer: Concepts, Measures, and Comparisons*, New York: Praeger, pp. 222–253.
Peck, Merton and Tamura, Shuji (1976) "Technology," in Hugh Patrick and Henry Rosovsky (eds) *Asia's New Giant*, Washington, DC: Brookings Institutions, pp. 525–585.
Roett, Riordan (1978) *Brazil: Politics in a Patrimonial Society*, New York: Praeger.

Rosenberg, Nathan (1972) *Technology and American Economic Growth*, White Plains, NY: M. E. Sharpe Inc.
Segal, Gary L. (1981) *Immigrating to Canada*, Vancouver, BC: Self-Counsel Press.
World Bank (1988) *China: Foreign Trade and Direct Foreign Investment*, Washington, DC: World Bank.
Yoshino, M. Y. (1975) "Japan as Host to the International Corporation," in Isaiah Frank (ed.) *The Japanese Economy in International Perspective*, Baltimore: The Johns Hopkins University Press, pp. 273–290.

2 ASSESSING JAPANESE TECHNOLOGY TRANSFER TO SOUTHEAST ASIA

Chew, Soon-Beng, Chew, Rosalind and Chan, Francis K. (1992) "Technology Transfer from Japan to ASEAN: Trends and Prospects," in Shojiro Tokunaga (ed.) *Japan's Foreign Investment and Asian Economic Interdependence: Production, Trade, and Financial Systems*, Tokyo: University of Tokyo Press, pp. 111–134.
Export-Import Bank of Japan (EXIM) (1989) *The Export-Import Bank of Japan: Role and Function*, Tokyo: Ex-Im Bank.
Hieneman, Bruce D., Johnson, Charles, Pamani, Ashok and Park, Hun Joon (1985) "Technology Transfer from Japan to Southeast Asia," in A. C. Samli (ed.) *Technology Transfer: Geographic, Economic, Cultural, and Technical Dimension*, Westport, Conn.: Quorum Books, pp. 143–153.
Ishi, Hiromitsu (1989) *The Japanese Tax System*, Oxford, England/New York: Clarendon Press.
Kawabe, Nobuo (1991) "Problems of the Perspectives on Japanese Management in Malaysia," in Shoichi Yamashita (ed.) *Transfer of Japanese Technology and Management to the ASEAN Countries*, Tokyo: University of Tokyo Press, pp. 239–266.
Kimbara, Tatsuo (1991) "Localization and Performance of Japanese Operations in Malaysia and Singapore," in Shoichi Yamashita (ed.) *Transfer of Japanese Technology and Management to the ASEAN Countries*, Tokyo: University of Tokyo Press, pp. 153–168.
Management and Coordination Agency (MCA) (1993) *Japanese Statistical Yearbook*, Tokyo: Sorifu Tokeikyoku.
Overseas Economic Cooperation Fund (OECF) (1990) *Annual Report 1990*, Tokyo: OECF.
Ozawa, Terutomo (1981) *Transfer of Technology from Japan to Developing Countries*, New York: UNITAR.
Phongpaichit, Pasuk (1991) "Japan's Investment and Local Capital in ASEAN since 1985," in Shoichi Yamashita (ed.) *Transfer of Japanese Technology and Management to the ASEAN Countries*, Tokyo: University of Tokyo Press, pp. 23–50.
Price Waterhouse (1989) *Doing Business in Indonesia*, Jakarta: PW.
Ramstetter, Eric D. (ed.) (1991) *Direct Foreign Investment in Asia's Developing Economies and Structural Change in the Asia–Pacific Region*, Boulder, Colo.: Westview Press.
Sim, A. B. (1978) *Decentralization and Performance: A Comparative Study*

of Malaysian Subsidiaries of Different National Origins, Kuala Lumpur: University of Malaysia.

Tokunaga, Shojiro (1992) "Japan's FDI-Promoting Systems and Intra-Asia Networks: New Investment and Trade Systems Created by the Borderless Economy," in Shojiro Tokunaga (ed.) *Japan's Foreign Investment and Asian Economic Interdependence: Production, Trade, and Financial Systems*, Tokyo: University of Tokyo Press, pp. 5–48.

Tokunaga, Shojiro (ed.) (1992) *Japan's Foreign Investment and Asian Economic Interdependence: Production, Trade, and Financial Systems*, Tokyo: University of Tokyo Press.

Urata, Shujiro (1991) "The Rapid Increase of Direct Investment Abroad and Structural Change in Japan," in Eric D. Ramstetter (ed.) *Direct Foreign Investment in Asia's Developing Economies and Structual Change in the Asia–Pacific Region*, Boulder, Colo.: Westview Press, pp. 175–199.

Yamashita, Shoichi (ed.) (1991) *Transfer of Japanese Technology and Management to the ASEAN Countries*, Tokyo: University of Tokyo Press.

Yonekawa, Shin'ichi (ed.) (1990) *General Trading Companies*, Tokyo: The United Nations University.

3 COMPARATIVE HOST GOVERNMENTS' FDI POLICIES

Behrman, Jack N. and Grosse, Robert E. (1990) *International Business and Governments: Issues and Institutions*, Columbia, SC: University of South Carolina Press.

Business International Corporation (BIC) (1987) *Investing, Licensing and Trading Conditions (India Section)*, New York: Business International Corporation.

Casson, Mark (1979) *Alternatives to the Multinational Enterprise*, London: Macmillan.

Chai, Denise (1993) "Tumen River Project: Boondoggle or Bonanza?," *Business Korea* (August): 25–29.

Graham, Edward M. and Krugman, Paul R. (1991) *Foreign Direct Investment in the United States*, Washington, DC: Institute for International Economics.

Hill, Charles W. L. (1994) *International Business: Competing in the Global Marketplace*, Burr Ridge, Ill.: Irwin.

Hood, S. and Young, S. (1979) *The Economics of the Multinational Enterprise*, London: Longman.

Jackson, James K. (1991) "Foreign Direct Investment in the United States," *CRS Issue Brief*, Washington, DC: Congressional Research Service.

Kleinberg, Robert (1990) *China's "Opening" to the Outside World*, Boulder, Colo.: Westview Press.

Lall, Sanjaya (1985) "Trade in Technology by a Slowly Industrializing Country: India," in Nathan Rosenberg and Claudio Frischtak (eds) *International Technology Transfer: Concepts, Measures, and Comparisons*, New York: Praeger, pp. 45–75.

Reich, Robert B. (1991) *The Work of Nations: Preparing Ourselves for the 21st Century*, New York: Alfred A. Knopf.

Thomsen, Stephen and Woolcock, Stephen (1993) *Direct Investment and European Integration*, New York: Council on Foreign Relations Press.
Todaro, Michael D. (1989) *Economic Development in the Third World*, New York: Longman, fourth edition.
Tolchin, Martin and Tolchin, Susan (1988) *Buying into America: How Foreign Money is Changing the Face of Our Nation*, New York: Times Books.
Villarreal, M. Angeles (1991) "Mexico's Maquiladora Industry," *CRS Report for Congress*, September 27, Washington, DC:, Congressional Research Service.
——— (1993) "Mexico's Changing Policy Toward Foreign Investment: NAFTA Implications," *CRS Report for Congress*, July 19, Washington, DC: Congressional Research Service.
Williams, Allan M. (1987) *The Western European Economy: A Geography of Post-war Development*, London: Hutchinson.
Woolcock, S. (1992) *Trading Partners or Trading Blows?*, New York: Council on Foreign Relations Press.

4 NATIONAL CONTROL OF TECHNOLOGY EXPORTS: THE CASE OF THE UNITED STATES

Berman, Harold J. and Garson, John R. (1967) "United States Export Controls – Past, Present, and Future," *Columbia Law Review* 67(5): 791–890.
Bertsch, Gary K. (1981) "US Export Controls: the 1970s and Beyond," *Journal of World Trade Law* 15(1): 67–82.
Bingham, Jonathan B. and Johnson, Victor C. (1979) "A Rational Approach to Export Controls," *Foreign Affairs* 57(4): 894–920.
Blakeney, Michael (1989) *Legal Aspects of the Transfer of Technology to Developing Countries*, Oxford: ESC Publishing Ltd.
Bruce, Peter and Fleming, Stewart (1987) "US Senate Votes to Ban Imports of Toshiba Products," *Financial Times* (July 2): 1.
Friedman, Thomas L. (1994) "US Ending Curbs on High-tech Gear to Cold War Foes," *New York Times* (March 30): 1.
Good, Alexander (1991) "The Changing Nature of United States Government Policy on the Transfer of Strategic Technology: An Overview," in Tamir Agmon and Mary Ann Von Glinow (eds) *Technology Transfer in International Business*, New York: Oxford University Press, pp. 37–55.
Hunt, Cecil (1983) "Multilateral Cooperation in Export Controls – the Role of COCOM," *University of Toledo Law Review* 14: 1285–1297.
Jacobsen, Hanns-Dieter (1985) "US Export Control and Export Administration Legislation," in Reinhard Rode and Hanns-D. Jacobsen (eds) *Economic Warfare or Detente*, Boulder, Colo.: Westview Press, pp. 213–225.
Lindell, Erik (1986) "Foreign Policy Export Controls," *California Management Review* 28(4): 27–39.
Long, William J. (1989) *US Export Control Policy: Executive Autonomy vs Congressional Reform*, New York: Columbia University Press.
Luks, Harold Paul (1987) "US National Security Export Controls: Legislative and Regulatory Proposals," *Balancing the National Interest: Working Papers*, Washington, DC: National Academy Press.

Luo, Wei (1994) *A Pathfinder to US Export Control Laws and Regulations*, Buffalo, NY: Williams S. Hein & Co. Inc.
Macdonald, Stuart (1990) *Technology and the Tyranny of Export Controls: Whisper Who Dares*, New York: St Martin's Press.
Melvern, Linda, Hebditch, David and Anning, Nick (1984) *Techno-Bandits*, Boston: Houghton Mifflin.
Richards, Eric L. (1994) *Law for Global Business*, Burr Ridge, Ill.: Irwin, Inc.
Smolenski, Mary (1992) "A Guide to Licensing and Shipping Dual-Use Products," *Business America* (February 10): 20–22.
Wrubel, Wende (1989) "The Toshiba–Kongsberg Incident: Shortcomings of COCOM, and Recommendations for Increased Effectiveness of Export Controls to the East Bloc," *American University Journal of International Law and Policy* 4: 241–261.

5 THE INTERNATIONAL DIMENSION OF INTELLECTUAL PROPERTY

Baxter, J. and Sinnott, J. (1983) *World Patent Law and Practice*, London: Sweet & Maxwell.
Behringer, John W. (1994) "Foreign Patents: Timing Is Everything," *Management Review* (April): 58–59.
Blakeney, Michael (1989) *Legal Aspects of the Transfer of Technology to Developing Countries*, Oxford: ESC Publishing Limited.
Chemical Marketing Reporter (*CMR*) (1992) "Biotech License Pact Worries US Industry" (December 28): 3.
Curesky, Karen (1989) "International Patent Harmonization Through WIPO," *Law and Policy in International Business* 21(2): 289–308.
Ewer, Sid R. (1993) "Protections of Trade Secrets and Intellectual Property," *Internal Auditor* (February): 46–48.
Hotchkiss, Carolyn (1994) *International Law for Business*, New York: McGraw-Hill, Inc.
Kotabe, Masaaki (1992) "A Comparative Study of US and Japanese Patent Systems," *Journal of International Business Studies* (First Quarter): 145–168.
Leaffer, Marshall A. (1990) *International Treaties on Intellectual Property*, Washington, DC: The Bureau of National Affairs, Inc.
O'Connor, Greg (1993) "Intellectual Property Rights and the Single Market: Good Intentions, Mixed Results, and Worrying Trend," *Business America* (March 8): 36–38.
Patent and Trademark Office of the US Department of Commerce (PTOUS) (1991) "An Introductory Guide for US Businesses on Protecting Intellectual Property Abroad," *Business America* (July 1): 2–7.
Robinson, Richard D. (1988) *The International Transfer of Technology: Theory, Issues, and Practice*, Cambridge, Mass.: Ballinger Publishing Company.
Twinomukunzi, C. (1982) "The International Patent System: A Third World Perspective," *Indian Journal of International Law* 22: 31–68.
United Nations Conference on Trade and Development (UNCTAD) (1975)

The Role of the Patent System in the Transfer of Technology to Developing Countries, Geneva: UNCTAD.

United Nations General Secretary (1972) *The Role of Patents in the Transfer of Technology to Developing Countries*, New York: United Nations.

World Intellectual Property Organization (WIPO) (1970) *Model Law for Developing Countries on Industrial Designs*, Geneva: WIPO.

——— (WIPO) (1979) *Model Law for Developing Countries on Inventions*, Geneva: WIPO.

——— (WIPO) (1988) *Background Reading on Intellectual Property*, Geneva: WIPO.

6 INTERNATIONAL PROTECTION OF INTELLECTUAL PROPERTY

Alikhan, Shahid (1993) "Intellectual Property Rights: The Paris Convention and Developing Countries," *Journal of Scientific and Industrial Research* 52(4): 219–224.

Behringer, John W. (1994) "Foreign Patents: Timing is Everything," *Management Review* (April): 58–59.

Blakeney, Michael (1989) *Legal Aspects of the Transfer of Technology to Developing Countries*, Oxford: ESC Publishing Limited.

Burger, Peter (1988) "The Berne Convention: Its History and its Key Role in the Future," *The Journal of Law and Technology* 3(1): 1–69.

Gorlin, Jacques J. (1993) "Update on International Negotiations on Intellectual Property Rights," in Mitchel B. Wallerstein, Mary Ellen McGee and Roberta A. Schoen (eds) *Global Dimensions of Intellectual Property Rights in Science and Technology*, Washington, DC: National Science Press, pp. 175–182.

Hill, Eileen (1994) "Intellectual Property Rights," *Business America* (January): 10–11.

Kerever, Andre (1991) "The Universal Copyright Convention," *The UNESCO Courier* (June 1): 50.

Leaffer, Marshall A. (ed.) (1990) *International Treaties on Intellectual Property*, Washington, DC: The Bureau of National Affairs, Inc.

Schechter, Roger E. (1991) "Facilitating Trademark Registration Abroad: The Implications of US Registration of the Madrid Protocol," *George Washington Journal of International Law and Economics* 25(2): 419–446.

Sinnott, John (1993) "The Paris Convention: What Changes Will be Made and When?," *Managing Intellectual Property* 33: 31–36.

World Intellectual Property Organization (WIPO) (1981) "The Nice Agreement" (March), Doc. CTMC/11, Geneva: WIPO.

——— (1983a) "Madrid Agreement and Trademark Registration Treaty (TRT)" (September) Doc. ISIP/83/7, Geneva: WIPO.

——— (1983b) "The Role of Patent Information and Documentation" (October), Doc. PI 105, Geneva: WIPO.

——— (1987) *International Classification of Goods and Services for the Purposes of the Registration of Marks (Nice Classification)*, WIPO pub. No. 500.1(E), 500.2(E), Geneva: WIPO, fifth edition.

——— (1989) *International Classification for Industrial Designs (Locarno Classification)*, WIPO pub. No. 501(E), Geneva: WIPO, fifth edition.
——— (1991) "New Procedures Will Make it Easier to File and Process an International Patent Application," *Business America* (September 23): 23–24.
Yankey, G. S. (1987) *International Patents and Technology Transfer to Less Developed Countries*, Avebury: Gower.

7 THE ISSUE OF INTELLECTUAL PROPERTY PROTECTION IN DEVELOPING COUNTRIES

Brueckmann, W. (1990) "Intellectual Property Protection in the European Community," in F. W. Rushing and C. G. Brown (eds) (1990) *Intellectual Property Rights in Science, Technology and Economic Performance: International Comparisons*, Boulder, Colo.: Westview Press, pp. 291–310.
Frischtak, Claudio (1990) "The Protection of Intellectual Property Rights and Industrial Technology Development in Brazil," in F. W. Rushing and C. G. Brown (eds) (1990) *Intellectual Property Rights in Science, Technology and Economic Performance: International Comparisons*, Boulder, Colo.: Westview Press, pp. 62–98.
Kunz-Hallstein, H. P. (1989) "The US Proposal for a GATT Agreement on Intellectual Property and the Paris Convention for the Protection of Industrial Property," *Vanderbilt Journal of Transnational Law* 22(2): 265–284.
Mansfield, Edwin (1990) "Intellectual Property, Technology and Economic Development," in F. W. Rushing and C. G. Brown (eds) (1990) *Intellectual Property Rights in Science, Technology and Economic Performance: International Comparisons*, Boulder, Colo.: Westview Press, pp. 17–30.
Marcus, Keith E. (1993) "IPRs and the Uruguay Round," *Federal Reserve Bank of Kansas City Economic Review* (First Quarter): 11–25.
Penrose, Edith (1951) *The Economics of the International Patent System*, Baltimore: Johns Hopkins University Press.
Primo Braga, Carlos Alberto (1990) "The Developing Country Case For and Against Intellectual Property Protection," in Wolfgang E. Siebeck, Robert E. Evenson, William Lesser and Carlos A. Primo Braga (eds) *Strengthening Protection of Intellectual Property in Developing Countries*, Washington, DC: World Bank, pp. 69–87.
Rapp, Richard T. and Rosek, Richard P. (1990) *Benefits and Costs of Intellectual Property Protections in Developing Countries*, Washington, DC: National Economic Research Associates, Inc. (NERA) Mimeograph.
Richards, Timothy J. (1988) *"Brazil," in T. J. Richards and R. Michael Gadbaw (eds), Intellectual Property Rights: Global Consensus, Global Conflict?*, Boulder Colo.: Westview Press pp. 149–185.
Robinson, Richard D. (1988) *The International Transfer of Technology: Theory, Issues, and Practice*, Cambridge, Mass.: Ballinger Publishing Co.
Rushing, F. W. and Brown, C. G. (eds) (1990) *Intellectual Property Rights in Science, Technology and Economic Performance: International Comparisons*, Boulder, Colo.: Westview Press.

Turner, Roger (1988) "Brazil: A Practical Guide to Intellectual Property Protection," *Business America* (January 18): 14–17.
Turner, Roger and MacNamara, Laurie (1989) "Intellectual Property Rights Protection in South America," *Business America* (June 5): 15–16.
UNCTAD (1985) *Draft International Code of Conduct on the Transfer of Technology*, TD/CODE TOT/47 (June 20), Geneva: UNCTAD.
—— (1987) *Trade and Development Report*, Geneva: UNCTAD.
Vaitsos, Constantine (1972) "Patents Revisited: Their Function in Developing Countries," *The Journal of Development* (October): 71–97.
Vasconcellos, Eduardo and Pereira, Hilda Solome (1994) "Patents in the Pharmaceutical Sector: The Case of Developing Countries," in W. M. Hoffman, J. B. Kamm, R. E. Frederick and E. S. Petry, Jr (eds) *Emerging Global Business Ethics*, Westport, Conn.: Quorum Books, pp. 234–241.

8 ENFORCING INTELLECTUAL PROPERTY PROTECTION

Baker, Lynn S. (1990) "Customs Cracks Down on International Thievery," *Global Trade* (April): 38–39.
Bello, Judith H. and Holmer, Alan F. (1990) "Special 301," *Fordham International Law Journal* 14(3): 874–880.
Brooks, Russell E. and Gellman, Gila E. (1993) "Combating Counterfeiting," *Marketing Management* 2(3): 49–51.
Feinberg, Robert M. and Rousslang, Ronald J. (1990) "The Economic Effects of IPR Infringements," *Journal of Business* 63(1): 79–89.
Hill, Eileen (1994a) "Intellectual Property Rights," *Business America* (January): 10–11.
——— (1994b) "Strong IPR Protection is Important for High-Tech Trade," *Business America* (August): 23–25.
Hotchkiss, Carolyn (1994) *International Law for Business*, New York: McGraw-Hill, Inc.
Lansing, Paul and Gabriella, Joseph (1993) "Clarifying Gray Market Gray Areas," *American Business Law Journal* 31(2): 313–337.
Mutti, John (1993) "Intellectual Property Protection in the United States under Section 337," *World Economy* 16(3): 339–357.
Patent and Trademark Office of the US Department of Commerce (PTO) (1991) "An Introductory Guide for US Businesses on Protecting Intellectual Property Abroad," *Business America* (July 1): 2–7.
Richards, Eric L. (1994) *Law for Global Business*, Burr Ridge, Ill.: Irwin.
Riordan, Teresa (1994) "US, Japan In Accord On Patents," *New York Times* (August 17): C1.
Robinson, Richard D. (1988) *The International Transfer of Technology: Theory, Issues, and Practice*, Cambridge, Mass.: Ballinger Publishing Company.
Simone, Joseph T. Jr, (1992) "Improving Protection of Intellectual Property," *The Chinese Business Review* (March–April): 9–11.
Smith, Guy C. (1988) "Protecting Intellectual Property Rights Abroad: Pointers for US Exporters," *Business America* (November 21): 14–15.

Svernlov, Carl (1992) "Super 301: Gone But Not Forgotten," *Journal of World Trade* 26(3): 125–132.

Tulchin, David, Klapper, Richard and Montagu, Alexandre (1994) "Lanham Act Review," *Managing Intellectual Property* 4: 17–19.

Weiser, Stuart M. (1993) "The NAFTA: A Watershed for Protection of Intellectual Property," *The International Lawyer* 27(3): 671–689.

Wineburg, Arthur and Jarbovsky, Isaac (1994) "New IPR Rules Can Help (Or Hurt) US Companies," *Foreign Trade* (September): 32–33.

Yosher, Shira R. (1992) "Competing in the Shadowy Gray: Protecting Domestic Trademark Holders from Gray Marketeers under the Lanham Act," *University of Chicago Law Review* 59(3): 1363–1390.

9 INTERNATIONAL LICENSING

Branch, Alan B. (1990) *Elements of Export Marketing and Management*, London: Chapman & Hall, second edition.

Buckley, Peter J. and Casson, Mark (1976) *The Future of the Multinational Enterprise*, London: Macmillan.

Casson, M. (1979) *Alternatives to the Multinational Enterprise*, New York: Holmes & Meier.

Caves, R. (1971) "International Corporations: The Industrial Economics of Foreign Investment," *Economica* 38: 1–27.

Contractor, Farok J. (1985) "Licensing Versus Foreign Direct Investment in US Corporate Strategy: An Analysis of Aggregate US Data," in Nathan Rosenberg and Claudio Frischtak (eds) *International Technology Transfer: Concepts, Measures and Comparisons*, New York: Praeger, pp. 277–320.

Davidson, William H. (1982) *Global Strategic Management*, New York: John Wiley & Sons.

Ehrbar, Thomas J. (1993) *Business International's Guide to International Licensing*, New York: McGraw-Hill.

James, Jr, Harvey S. and Weidenbaum, Murray (1993) *When Businesses Cross International Borders: Strategic Alliances and Their Alternatives*, Westport, Conn.: Praeger.

Killing, J. Peter (1980) "Technology Acquisition: License Agreement or Joint Venture," *Columbia Journal of World Business* 15(3): 38–46.

Langenecker, Juliane (1993) "Licensing Lives On," *Business Eastern Europe* (September 27): 1–2.

Lineback, J. Robert (1994) "New Licensing Demands Coming from TI," *Electronic Business Buyer* (June): 26.

Mowery, David C. (1988) "An Overview," in David C. Mowery (ed.) *International Collaborative Ventures in US Manufacturing*, Cambridge, Mass.: Ballinger Publishing Co., pp. 1–22.

Root, Franklin R. (1994) *Entry Strategies for International Markets*, New York: Lexington Books.

Rugman, Alan M., Lecraw, Donald J. and Booth, Laurence D. (1985) *International Business*, New York: McGraw-Hill.

Shahrokhi, Manuchehr (1987) *Reverse Licensing: International Technology Transfer to the United States*, Westport, Conn.: Praeger.

Sherman, Andrew J. (1992) "Technology License Agreements," *D&B Reports* (May/June): 42.
Teece, David J. (1977) *The Multinational Corporation and the Resource Cost of International Technology Transfer*, Cambridge, Mass.: Ballinger.
——— (1981) "The Market for Know-how and the Efficient International Transfer of Technology," *Annals of the American Academy of Political and Social Science* 458: 81–96.
Telesio, Piero (1979) *Technology Licensing and Multinational Enterprises*, New York: Praeger.

10 PRICE NEGOTIATIONS IN INTERNATIONAL TECHNOLOGY LICENSING

Arni, Venkata R. S. (1984) *Evaluation of Technology Payments*, UNIDO, I.D./W.G. 429/5 (September 6), Vienna: UNIDO.
Bidault, Francis (1989) *Technology Pricing: From Principles to Strategy*, New York: St Martin's Press.
Bingham, Frank G., Jr and Faffield, Barney T., III (1990) *Business to Business Marketing Management*, Homewood, Ill.: Richard D. Irwin.
Buckley, P. and Casson, M. (1976) *The Future of the Multinational Enterprise*, New York: Holmes & Meier.
Cho, K. R. (1988) "Issues of Compensation in International Technology Licensing," *Management International Review* 28(2): 70–79.
Contractor, Farok J. (1980) "'Profitability' of Technology Licensing by US Multinationals," *Journal of International Business Studies* (Fall): 40–63.
——— (1981) *International Technology Licensing*, Lexington, Mass.: D. C. Heath and Company.
——— (1985) *Licensing in International Strategy: A Guide for Planning and Negotiations*, Westport, Conn.: Quorum Books.
——— and Sagafi-nejad, T. (1981) "International Technology Transfer: Major Issues and Policy Responses," *Journal of International Business Studies* (Fall): 113–135.
Davies, H. (1977) "Technology Transfer Through Commercial Transactions," *Journal of Industrial Economics* 26(2): 161–191.
Friedman, David D. (1990) *Price Theory*, Cincinnati, OH: South-Western Publishing Co.
Kotler, Philip (1991) *Marketing Management: Analysis, Planning, Implementation, and Control*, Englewood Cliffs, NJ: Prentice-Hall, seventh edition.
Li, Linhuan (1989) *International Technology Transfer and Acquisition*, Beijing: Employee Education Press.
Nagle, Thomas T. (1987) *The Strategy and Tactics of Pricing: A Guide to Profitable Decision Making,* Englewood Cliffs, NJ: Prentice-Hall Inc.
Root, Franklin R. (1981) "The Pricing of International Technology Transfers via Non-Affiliate Licensing Arrangements," in Tagi Sagafi-nejad, Richard W. Moxon and Howard V. Perlmutter (eds) *Controlling International Technology Transfer: Issues, Perspectives, and Policy Implications*, New York: Pergamon Press, pp. 120–133.
——— and Contractor, Farok J. (1981) "Negotiating Compensation in

International Licensing Agreements," *Sloan Management Review* (Winter): 23–32.

United Nations Industrial Development Organization (UNIDO) (1983) *Technology Payments Evaluation: Summary Results of a Pilot Exercise*, Caracas (October 17–20), Vienna: UNIDO.

Watkins, William M. (1990) *Business Aspects of Technology Transfer: Marketing and Acquisition*, Park Ridge, NJ: Noyes Publications.

Yin, Zun-sheng (1993) *The Price of International Technology Transfer*, Shanghai: People's Press.

11 LICENSING AGREEMENT

Arnold, Tom (1984) *Domestic and Foreign Technology Licensing*, New York: Practicing Law Institute.

Bidault, Francis (1989) *Technology Pricing: From Principles to Strategy*, New York: St Martin's Press.

Brookhart, Walter R., Leach, Sidney M. and Tobor, Ben D. (1980) *Current International Legal Aspects of Licensing and Intellectual Property*, Chicago: American Bar Association.

Contractor, Farok J. (1981) *International Technology Licensing: Compensation, Costs, and Negotiation*, Lexington, Mass.: Lexington Books.

——— (1985) *Licensing in International Strategy: A Guide for Planning and Negotiations*, Westport, Conn.: Quorum Books.

Ehrbar, Thomas J. (1993) *Business International's Guide to International Licensing*, New York: McGraw-Hill, Inc.

Finnegan, Marcus B. (1980) "A Code of Conduct Regulating International Technology Transfer: Panacea or Pitfall?," in Walter R. Brookhart, S. M. Leach and B. D. Tobor (eds) *Current International Legal Aspects of Licensing and Intellectual Property*, Chicago: American Bar Association, pp. 38–110.

Richards, Eric L. (1994) *Law for Global Business*, Burr Ridge, Ill.: Irwin.

Robinson, Richard D. (1988a) *Cases on International Technology Transfer*, Gig Harbor, WA: Hamlin Publications.

——— (1988b) *Teaching Notes for Cases on International Technology Transfer*, Gig Harbor, WA: Hamlin Publications.

Root, Franklin R. (1994) *Entry Strategies for International Markets*, New York: Lexington Books.

Shahrokhi, Manuchehr (1987) *Reverse Licensing: International Technology Transfer to the United States*, Westport, Conn.: Praeger.

Sherman, Andrew J. (1991) *Franchising and Licensing: Two Ways to Build Your Business*, New York: American Management Association.

United Nations (1973) *Guide for the Acquisition of Foreign Technology in Developing Countries, with Special Reference to Technology Agreements*, New York: UN.

United Nations Economic Commission for Europe (UNECE) (1980) *Manual on Licensing Procedures in Member Countries of United Nations Economic Commission for Europe*, New York: Clark Boardman Co.

United Nations Industrial Development Organization (UNIDO) (1989)

Guide to Guarantee and Warranty Provisions in Transfer-of-Technology Transactions, Vienna: UNIDO.

Van Horn, Mike (1989) *Pacific Rim Trade: The Definitive Guide to Exporting and Investment*, New York: AMACOM.

Watkins, William M. (1990) *Business Aspects of Technology Transfer: Marketing and Acquisition*, Park Ridge, NJ: Noyes Publications.

World Intellectual Property Organization (WIPO) (1977) *Licensing Guide for Developing Countries*, Geneva: WIPO.

12 THE ISSUE OF RESTRICTIVE BUSINESS PRACTICES IN TECHNOLOGY TRANSFER

Andean Pact, Common Regime of Treatment of Foreign Capital and of Trademarks, Patents, Licenses, and Royalties, Articles 18–26, 1971.

Blakeney, Michael (1989) *Legal Aspects of the Transfer of Technology to Developing Countries*, Oxford: ESC Publishing Limited.

Cabanellas, Guillermo Jr, (1984) *Studies in Industrial Property and Copyright Law*, Weinheim: Verlag Chemie.

Cheeseman, Henry R. (1992) *Business Law: The Legal, Ethical, and International Environment*, Englewood Cliffs, NJ: Prentice-Hall.

Finnegan, Marcus B. (1980) "A Code of Conduct Regulating International Technology Transfer: Panacea or Pitfall?," in W. R. Brookhart, S. M. Leach and B. D. Tobor (eds) *Current International Legal Aspects of Licensing and Intellectual Property*, Chicago: American Bar Association, pp. 38–110.

Goosen, Richard J. (1987) *Technology Transfer in the People's Republic of China: Law and Practice*, Dordrecht/Boston: Martinus Nijhoff Publishers.

Hirose, Hisao (1984) "Japan: the Prohibition of Cartels and Industrial Defenses Against It," in Julian Maitland-Walker (ed.) *International Antitrust Law, Vol. 1: A Review of National Laws*, Oxford: ESC Publishing Limited, pp. 102–113.

International Patent and Know-how Licensing Task Force (1981) *US Antitrust Law in International Patent and Know-how Licensing*, Chicago: American Bar Association.

Li, Shaoqing (1993) *International Technology Transfer*, Guangzhou: Jinan University Press.

Long, Frank (1981) *Restrictive Business Practice, Transnational Corporations, and Development: A Survey*, Dordrecht/Boston: Martinus Nijhoff Publishers.

Maitland-Walker, Julian (ed.) (1984) *International Antitrust Law, Vol. 1: A Review of National Laws*, Oxford: ESC Publishing Limited.

Marks, David H. (1984) "United States of America: Antitrust Enforcement in the US," in Julian Maitland-Walker (ed.) *International Antitrust Law, Vol. 1: A Review of National Laws*, Oxford: ESC Publishing Limited, pp. 156–178.

Marks, Julius J. and Samie, Najeeb (1983) *Antitrust and Restrictive Business Practices: International, Regional and National Regulation*, New York: Oceana Publications, Inc.

Matsushita, Mitsuo and Schoenbaum, Thomas J. (eds) (1989) *Japanese*

International Trade and Investment Law, Tokyo: University of Tokyo Press.
Regulations to the Law on Control and Registration of Technology Transfer and Use and Exploitation of Patents and Trademarks, effective on January 10, 1990 (Mexican Law).
Roffe, Pedro (1985) "Transfer of Technology: UNCTAD's Draft International Code of Conduct," *The International Lawyer* 19: 689–695.
United Nations (1948) *The Havana Charter*, New York: United Nations.
United Nations Conference on Trade and Development (UNCTAD) (1985) *Draft International Code on the Transfer of Technology*, TD/CODE TOT/47 (June 20), Geneva: UNCTAD.
——— (1987) "Negotiations on a Draft International Code of Conduct on the Transfer of Technology," UNCTAD doc. TD/CODE TOT/51 (October 21), Geneva: UNCTAD.
World Intellectual Property Organization (WIPO) (1977) *Licensing Guide for Developing Countries*, Geneva: WIPO.

13 OTHER MAJOR COMMERCIAL CHANNELS FOR TECHNOLOGY TRANSFER

Blakeney, Michael (1989) *Legal Aspects of the Transfer of Technology to Developing Countries*, Oxford: ESC Publishing Ltd.
Cavusgil, S. Tamer (1985) "Multinational Corporations and the Management of Technology Transfers," in A. Coskun Samli (ed.) *Technology Transfer: Geographic, Economic, Cultural, and Technical Dimensions*, Westport, Conn.: Quorum Books.
Chan, Peng S. and Justis, Robert T. (1990) "Franchise Management in East Asia," *Academy of Management Executives* 4(2): 75–85.
Harrigan, Kathryn R. (1986) *Managing for Joint Venture Success*, Lexington, Mass.: Lexington Books.
Hladik, Karen J. (1988) "R&D and International Joint Ventures," in Farok J. Contractor and Peter Lorange (eds) *Cooperative Strategies in International Business*, Lexington, Mass.: Lexington Books, pp. 187–203.
James Jr, Harvey S. and Weidenbaum, Murray (1993) *When Businesses Cross International Borders: Strategic Alliances and Their Alternatives*, Westport, Conn.: Praeger.
Kaynak, Erdener and Dalgic, Tevfic (1992) "Internationalization of Turkish Construction Companies: A Lesson for Third World Countries?," *Columbia Journal of World Business* 26(4): 60–75.
Keller, Robert T. (1990) "International Technology Transfer: Strategies for Success," *Academy of Management Executive* 4(2): 33–43.
Kotabe, Masaaki (1989) "'Hollowing-out' of US Multinationals and Their Global Competitiveness: An Intrafirm Perspective," *Journal of Business Research* (August): 1–16.
Lewis, Jordan (1990) *Partnerships for Profit: Structuring and Managing Strategic Alliances*, New York: Free Press.
Livingstone, J. M. (1989) *The Internationalization of Business*, New York: St Martin's Press.
McVey, Thomas B. (1985) "Overview of the Commercial Practice of

Countertrade," in Bart S. Fisher and Kathleen M. Harte (eds) *Barter in the World Economy*, New York: Praeger, pp. 9–35.

Marion, Larry (1994) "The Hidden Growth of the Contract Manufacturing," *Electronic Business Buyer* (August): 54.

Root, Franklin R. (1994) *Entry Strategies for International Markets*, New York: Lexington Books.

Schaffer, Matt (1989) *Winning the Countertrade War: New Export Strategies for America*, New York: John Wiley & Son.

Sherman, Andrew J. (1991) *Franchising and Licensing*, New York: AMACOM.

Spiegel, Yossef (1993) "Horizontal Subcontracting," *Rand Journal of Economics* 24(4): 570–590.

Steinberg, Carol (1993) "Mating Game," *World Trade* (April): 147–158.

14 INTERNATIONAL SUBCONTRACTING AND SELECTED INDUSTRIES

Avery, Susan (1993) "Contract Manufacturing Is Up," *Purchasing* (November 11): 57.

AW&ST (1991) "Boeing Signs Three Subcontractors for Major 777 Component Work," *Aviation Week & Space Technology* (AW&ST) (March 25): 41.

Bamberg, G. and Spremann, K. (eds) (1987) *Agency Theory, Information, and Incentives*, Berlin and New York: Springer-Verlag.

Dicken, P. (1986) *Global Shift, Industrial Change in a Turbulent World*, London: Harper & Row.

Fagan, Mark L. (1991) "A Guide to Global Sourcing," *The Journal of Business Strategy* (March/April): 21–25.

Germidis, Dimitri (ed.) (1980) *International Subcontracting: A New Form of Investment*, Paris: Development Center of OECD.

Helper, Susan (1991) "How Much Has Really Changed Between US Automakers and Their Suppliers?," *Sloan Management Review* (Summer): 15–28.

Hibbert, E. P. (1993) "Global Make-or-Buy Decisions," *Industrial Marketing Management* 22: 67–77.

Houlder, Vanessa (1995) "Revolution in Outsourcing," *Financial Times* (January 6): 14.

Jansson, H. (1982) *Interfirm Linkages in a Developing Economy: The Case of Swedish Firms in India*, Uppsala, Sweden: Uppsala University.

Marion, Larry (1994) "The Hidden Growth of the Contract Manufacturing," *Electronic Business Buyer* (August): 54.

Michalet, Charles-Albert (1980) "International Sub-contracting: A State-of-the-Art," in Dimitri Germidis (ed.) *International Subcontracting: A New Form of Investment*, Paris: Development Center of OECD, pp. 38–70.

Ristelhueber, Robert (1994) "Contract Manufacturing: Outsources Becomes Part of the Team," *Electronic Business Buyer* (August): 48–63.

Sang, Chul Suh (1975) "Development of a New Industry Through Exports: The Electronics Industry in Korea," in Wontack Hong and Anne O.

Krueger (eds) *Trade and Development in Korea*, Seoul: Korea Development Institute, pp. 118–122.
Tully, Shawn (1994) "You'll Never Guess Who Really Makes . . . ," *Fortune* (October 3): 124–128.
United Nations Industrial Development Organization (UNIDO) (1974) *Subcontracting for Modernizing Economies,* New York: United Nations Press.
Watanabe, Susumu (1980) "International Subcontracting and Regional Economic Integration of the ASEAN Countries: The Roles of Multinationals," in Dimitri Germidis (ed.) *International Subcontracting: A New Form of Investment*, Paris: Development Center of OECD, pp. 210–227.
Williamson, O. E. (1986) *Economic Organization: Firms, Markets and Policy Control*, Brighton, Sussex: Wheatsheaf Books.
Wong, Poh Kam (1991) *Technological Development Through Subcontracting Linkages*, Singapore: Asian Productivity Organization.

15 TRANSFERRING TECHNOLOGY TO CHINA VIA JOINT VENTURES

Casati, Christine (1991) "Satisfying Labor Laws – and Needs," *The China Business Review* (July–August): 16–22.
Chen, Min (1993) "Tricks of the China Trade," *The China Business Review* (March–April): 12–16.
Cohen, Jerome Alan and Pierce, David G. (1987) "Legal Aspects of Licensing Technology," *The China Business Review* (May–June): 44–49.
De Bruijn, Erik J. and Jia, Xiangfeng (1993) "Transferring Technology to China by Means of Joint Ventures," *Research-Technology Management* (January–February): 17–22.
Epser, Pavel (1991) "Managing Chinese Employees," *The China Business Review* (July–August): 24–30.
Fischer, Hans H. (1993) "Joint Venturing in China: A Negotiating Experience," *East Asian Executive Reports* (May): 9–16.
Goldenberg, Susan (1988) *Hands Across the Ocean: Managing Joint Ventures*, Boston: Harvard Business School Press.
Goosen, Richard J. (1987) *Technology Transfer in the People's Republic of China: Law and Practices,* Dordrecht/Boston: Martinus Nijhoff Publishers.
Hendryx, Steven R. (1986) "Implementation of a Technology Transfer Joint Venture in the People's Republic of China: A Management Perspective," *Columbia Journal of World Business* (Spring): 57–66.
Ho, Alfred K. (1990) *Joint Ventures in the People's Republic of China*, New York: Praeger.
Ireland, Jill (1991) "Finding the Right Management Approach," *The China Business Review* (January–February): 14–17.
Mann, Jim (1989) *Beijing Jeep,* New York: Simon & Schuster.
Peck, Joyce (1991) "Standardizing Foreign Income Taxes," *The China Business Review* (September–October): 12–15.
Stelzer, Leigh, Chunguang, Ma and Banthin, Joanna (1991) "Gauging Investor Satisfaction," *The China Business Review* (November–December): 54–56.

The Income Tax Law of the People's Republic of China for Enterprises with Foreign Investment and Foreign Enterprises ("The United Income Tax Law"), April 9, 1991, Beijing.

The Law of the People's Republic of China on Joint Ventures Using Chinese and Foreign Investment ("The Law"), July 1, 1979, Beijing.

The Provisional Regulations for Exchange Control of the People's Republic of China ("The Provisional Regulations"), December 1980, Beijing.

The Regulations for the Implementation of the Law of the People's Republic of China on Joint Ventures Using Chinese and Foreign Investment ("The Regulations"), September 20, 1983, Beijing.

The Regulations on Labor Management in Joint Venture Using Chinese and Foreign Investment ("The Regulations on Labor Management"), July 26, 1980, Beijing.

16 ISSUES IN DISPUTE RESOLUTION

Barnett, William (1994) "Are Banks Missing the Boat on Alternative Dispute Resolution Services?," *Banking Law Journal* 111(3): 309–313.

Berman, Peter J. (1994) "Resolving Business Disputes Through Mediation and Arbitration," *CPA Journal* 64(11): 74–77.

Collier, J. G. (1987) *Conflict of Law,* Cambridge, England: Cambridge University Press.

Craig, W. L., Park, W. W. and Paulsson, J. (eds) (1985) *International Chamber of Commerce Arbitration*, New York: Oceana Publications.

Dore, Isaak I. (1986) *Arbitration and Conciliation Under the UNCITRAL*, Boston: Martinus Nijhoff Publishers.

Eisenstadt, S.N. and Ben-Ari, Eyal (eds) (1990) *Japanese Models of Conflict Resolution*, London: Kegan Paul International.

Fitzmaurice, Gerald, Sir (1986) *The Law and Procedure of the International Court of Justice*, Cambridge, England: Grotius Publications.

Fox, William F. (1992) *International Commercial Agreements: A Primer on Drafting, Negotiating and Resolving Disputes*, Boston: Kluwer Law and Taxation Publishers.

Gurry, Francis (1994) "The WIPO Arbitration Center," *Managing Intellectual Property* 3(37): 4–6.

Handbook on the Operation of the Hague Service Convention of 1965 (Handbook) (1983), Antwerpen: Maarten Kluwer's International Uitgeversonderneming.

Hanlon, Michelle L. D. (1991) "The Japan Commercial Arbitration Association: Arbitration with the Flavor of Conciliation," *Law & Policy in International Business* 22(3): 603–626.

Hotchkiss, C. (1994) *International Law for Business*, New York: McGraw-Hill.

Mason, Paul E. (1994) "International Commercial Arbitration," *Dispute Resolution Journal* 49(2): 22–26.

Neale, A. D. (1988) *International Business and National Jurisdiction*, Oxford: Clarendon Press.

Nolan-Haley, Jacqueline M. (1992) *Alternative Dispute Resolution*, Minneapolis/St Paul: West Publishing Co.

Ratliff, Alan G. (1994) "To Litigate or Not to Litigate?," *Baylor Business Review* 12(1): 9–10.

Redfern, Alan and Hunter, Martin (1986) *Law and Practice of International Commercial Arbitration*, London: Sweet & Maxwell.

Richards, Eric L. (1994) *Law for Global Business*, Burr Ridge, Ill.: Irwin.

Schaffer, Richard, Earle, Beverley and Agusti, Filiberto (1993) *International Business Law and Its Environment*, Minneapolis/St Paul: West Publishing Co.

Van den Berg, Albert Jan (1981) *The New York Arbitration Convention of 1958*, Deventer/Boston: Kluwer Law and Taxation Publishers.

Von Glahn, Gerhard (1992) *Law Among Nations: An Introduction to Public International Law*, New York: Macmillan Pub. Co., sixth edition.

CONCLUDING REMARKS

Aggarwal, Raj (1991) "Technology Transfer and Economic Growth: A Historical Perspective on Current Developments," in Tamir Agmon and Mary Ann Von Glinow (eds) *Technology Transfer in International Business*, New York: Oxford University Press, pp. 56–76.

Agmon, Tamir and Von Glinow, Mary Ann (1991) "Conclusion," in Tamir Agmon and Mary Ann Von Glinow (eds) *Technology Transfer in International Business*, New York: Oxford University Press, pp. 273–274.

Asheghian, Parviz and Ebrahimi, Bahman (1990) *International Business: Economics, Environment and Strategies*, New York: HarperCollins Publishers.

Blakeney, Michael (1989) *Legal Aspects of the Transfer of Technology to Developing Countries*, Oxford: ESC Publishing Ltd.

Cho, K. R. (1988) "Issues of Compensation in International Technology Licensing," *Management International Review* 28(2): 70–79.

Contractor, Farok J. (1985) *Licensing in International Strategy: A Guide for Planning and Negotiations,* Westport, Conn.: Quorum Books.

Grosse, Robert and Kujawa, Duane (1995) *International Business: Theory and Managerial Applications*, Homewood, Ill.: Irwin.

Heller, Peter B. (1985) *Technology Transfer and Human Values: Concepts, Applications, Cases*, New York: Lanham.

Keller, Robert T. and Chinta, Ravi R. (1990) "International Technology Transfer: Strategies for Success," *Academy of Management Executive* 4(2): 33–43.

Perlmutter, H. V. and Sagafi-nejad, Tagi (1981) *International Technology Transfer*, New York: Pergamon Press.

Robinson, Richard D. (1988) *The International Transfer of Technology*, Cambridge, Mass.: Ballinger Publishing Co.

Robock, Stefan H. and Simmonds, Kenneth (1989) *International Business and Multinational Enterprises*, Homewood, Ill.: Irwin, fourth edition.

Root, Franklin R. (1981) "The Pricing of International Technology Transfers via Non-Affiliate Licensing Arrangements," in T. Sagafi-nejad, R. W. Moxon and H. V. Perlmutter (eds) *Controlling International Technology Transfer: Issues, Perspectives, and Policy Implications*, New York: Pergamon Press, pp. 120–133.

——— and Contractor, Farok J. (1981) "Negotiating Compensation in International Licensing Agreements," *Sloan Management Review* (Winter): 23–32.

Samli, A. C. (1985) "Technology Transfer: The General Model," in A. C. Samli (ed.) *Technology Transfer: Geographic, Economic, Cultural, and Technical Dimensions*, Westport, Conn.: Quorum Books, pp. 3–15.

Toyne, Brian (1989) "International Exchange: A Foundation for Theory Building in International Business," *Journal of International Business Studies* (Spring): 1–17.

Vaghefi, R. Reza, Paulson, Steven K. and Tomlinson, William H. (1991) *International Business: Theory and Practice*, New York: Taylor & Francis.

Index